Black-Box Models of Co
in Cryptology

Tibor Jager

Black-Box Models of Computation in Cryptology

Foreword by Prof. Dr. Jörg Schwenk

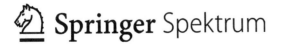 **Springer** Spektrum

RESEARCH

Tibor Jager
Velbert, Germany

Dissertation Ruhr-Universität Bochum, 2011

ISBN 978-3-8348-1989-5   ISBN 978-3-8348-1990-1 (eBook)
DOI 10.1007/978-3-8348-1990-1

The Deutsche Nationalbibliothek lists this publication in the Deutsche Nationalbibliografie;
detailed bibliographic data are available in the Internet at http://dnb.d-nb.de.

Springer Spektrum

Cover design: KünkelLopka GmbH, Heidelberg

Printed on acid-free paper

Springer Spektrum is a brand of Springer DE. Springer DE is part of Springer Science+Business Media.
www.springer-spektrum.de

Foreword

Theoretical Computer Science as a new research area was established by pioneers like Alan Turing (1912-1954) and Alonzo Church (1903-1995) around the question which mathematical functions were computable, and which not. The Church-Turing thesis states equivalence on the early notions of computability, which solves this first question. However, there are computable functions that grow so fast that in practice they cannot be computed (e.g. the Ackermann function).

After the first computers had been built, it soon became clear that the resources (computing time and storage space) needed to compute a function also limited the range of functions that could practically be computed. Systematic studies were initiated by the seminal paper of Juris Hartmanis and Richard Stearns in 1965. Shortly after, Stephen Cook and Leonid Levin independently proved that there exist practically important problems that most probably were impossible to solve on Turing-equivalent computers. These so-called NP-complete problems include e.g. the well-known Travelling Salesman problem. In 1972 Richard Karp described reductions between these problems, a technique that should become extremely important in modern cryptography.

Nobody could imagine at this time that problems which are impossible to solve given restrictions on computing time should be of any use. Thus the ideas of Whitfield Diffie and Martin Hellman, published in their seminal paper *"New Directions in Cryptography"* in 1976, aroused controversy (which finally resulted in the discovery of the RSA algorithm in 1978). Nevertheless, it laid the foundations for modern cryptography, which is one of the most important areas of applied mathematics today. The concept of "one-way functions" introduced in this paper can be found, with many sophisticated variations, in nearly every paper published in the area of Public-Key Cryptography.

For one-way functions, we need an asymmetry in computational complexity: It must be practically feasible to evaluate a function f at a given point, i.e. the computational resources needed to compute $f(x)$ must be bounded. On the other hand, computing the preimage of a function value should ideally need a practically unbounded amount of resources (at least computing time).

At a first glance, this concept very much resembles the complexity class NP ("Nondeterministc Polynomial Time"), which plays a central role in complexity theory. The famous assumption P $\neq$ NP implies the existence of one-way functions, but yet remains unproven. Moreover, this assumption only implies worst-case complexity, and only for NP-complete problems.

In Public-Key Cryptography, two mathematical structures play a fundamental role: commutative groups, and rings modulo a composite integer. In both structures, problems are known which are assumed to be practically uncomputable: the discrete logarithm problem in groups, and computing roots in these rings. Both problems are not NP-complete, and both require average-case intractability. So the challenge was to secure the foundations of modern cryptography by proving some lower bounds on the computational complexity of these problems. In the Turing machine model, this seems impossible: researchers tried to prove such lower bounds since the early seventies of the last century, and failed. It was thus necessary to restrict the computation model, and several restrictions had been tried in the past.

For the two algebraic problems mentioned above, Victor Shoup in 1997 most convincingly proposed an algebraic restriction, and supported this restriction with an impressive result: For solving the discrete logarithm problem in commutative groups, he restricted the operations of the "computer" to performing group operations and testing equality. He was able to show a lower bound on the time complexity of these generic algorithms, which exactly matched the complexity of known generic algorithms. His result implied that all improved algorithms (which exist for certain groups) must be non-generic.

The paper by Shoup was the starting point for research in the area of generic algorithms. This research soon included finite fields, and the above mentioned ring structures. Research in this area at the Horst Görtz Institute for IT Security (HGI) started independently from the established chairs, and was performed solely by postdocs and PhD students. Andy Rupp was the first to work in this field. Together with Gregor Leander he achieved the first important result by showing that in the generic ring model, the RSA problem is equivalent to factoring integers, thus solving a special case of one of the most important open questions in cryptography.

However, the elegance of this proof raised questions about the suitability of the generic model for indicating lower bounds also in the Turing machine model (where the RSA problem is still an open research issue). Up to this point, the generic model was used as a "test model" to see if an algebraic structure was suitable for cryptography. Thus e.g. new algorithms and assumptions based on bilinear pairings were checked in the generic model, before introducing them.

The author of the book presented here took up all these loose ends and stretched the generic model to its limits: Tibor Jager, who was introduced to this field by the two aforementioned researchers, and works in close collaboration with them, gives a unified treatment on all variants of generic models, and presents the most important proofs in a clear and readable fashion. His research is centered around three main questions:

- Are generic models a reasonable abstraction of reality?
- Can we weaken these models (i.e. give the computer more power) to better adapt these models?
- Is solving the Diffie-Hellman problem equivalent to solving the discrete logarithm problem?

Chapter 3 answers the third question in the negative, but only for algebraic rings modulo a composite integer. (The question remains open for algebraic groups of prime order.) This result was published together with Altmann and Rupp, but the proof given here is much simpler and elegant.

In Chapter 4, the first question is answered in the negative for generic ring algorithms as used by Leander and Rupp in 2006, and Aggarval and Maurer in 2009. This result clearly shows the limits of generic algorithms, thus marking a boundary of their applicability: In the generic ring model, both deciding quadratic residuosity and computing Jacobi symbols is infeasible. However, in the Turing machine model, Jacobi symbols are easy to compute (using the algorithm given by Gauss), whereas the quadratic residuosity problem is assumed to remain hard. This shows that results proved for generic ring algorithms cannot give any indication for Turing machine complexity, and should thus be treated with great care.

Semi-generic algorithms are proposed as a means to achive results with better applicability to the "real" world. The most prominent example for the applicability of such algorithms are bilinear pairings, where the constructions typically involve groups based on elliptic curves (where only generic algorithms are known) and prime order groups over the integers (where better, non-generic algorithms can be applied). By allowing non-generic algorithms in the proof for lower bounds where applicable, the semi-generic model is better adapted to the real problem than a purely generic approach. As a promising result Mr. Jager has shown that the Computational Diffie-Hellman problem (CDH), adapted to the bilinear setting, is semi-generically equivalent to the corresponding discrete logarithm problem.

To summarize: The book authored by Tibor Jager is the most comprehensive overview on a restricted computational model, the generic algorithm model, for algebraic groups, rings, fields, and other cryptographically relevant domains. It presents both the major achievements in this field in terms of lower computational bounds and equivalence results, and shows the limitations of this approach. By carefully considering these limitations, generic algorithms are one of the most important foundations for modern cryptography.

Prof. Dr. Jörg Schwenk

Acknowledgements

This book is based on a PhD thesis under the supervision of Prof. Dr. Jörg Schwenk, submitted to the Faculty of Electrical Engineering and Information Technology at Ruhr-University Bochum in July 2011. I am indebted to many people who helped me, either directly or indirectly, to finish this project. Unfortunately, I can name only a few here.

First and foremost I would like to thank Jörg Schwenk, for his support and advice, the joint work, and for many delightful conversations on cryptography and IT security. I am also very grateful to my second supervisor Dennis Hofheinz, for many helpful comments on this text and enlightening discussions on further cryptographic topics. Thanks to Andy Rupp for the pleasant and productive joint work on generic group and ring algorithms.

I would like to thank all members of the Chair for Network and Data Security, for the friendly working atmosphere and the good collaboration on research, teaching, and industrial projects. In particular I would like to thank Sven Schäge for mutually explaining many cryptographic techniques and discussing countless thoughts and ideas, and Florian Kohlar and Juraj Somorovsky for the joint work. Thanks also to Petra Winkel and Jürgen Weide for their help with administrative obstacles.

Thanks to Alex May and all members of the Chair for Cryptology and IT Security, for many pleasant and helpful discussions and for letting me participate in their research seminar. In particular I would like to mention Maike Ritzenhofen, Mathias Herrmann, and Alexander Meurer. I am furthermore very thankful to Eike Kiltz for explaining cryptographic concepts, discussing many ideas, and being my host when I visited the CWI Amsterdam in 2009.

Am meisten danke ich Anna, Stefan, Ulla und Atila, für die ständige und unbedingte Unterstützung.

Tibor Jager

Contents

1 Introduction

The security of many cryptosystems relies on assumptions that certain computational problems, mostly from number theory and algebra, are intractable. Therefore it is important to study the validity of these assumptions. Ideally, we would like to show that these assumptions hold in a standard model of computation, e.g. where algorithms intending to solve a computational problem are modeled as Turing machines with reasonably restricted running time. Unfortunately, proving useful lower complexity bounds in the standard model seems to be impossible with currently available techniques.

Many important hardness assumptions are based on computational problems defined over algebraic groups. Famous examples are, for instance, the discrete logarithm problem [DH76] and the RSA problem [RSA78]. A natural approach to analyze these assumptions is to consider algorithms solving a given problem by performing only the abstractly defined properties of a group, without exploiting specific properties of the representation of group elements. Indeed, there are group representations, such as for instance certain elliptic curve groups, for which only very few properties beyond the abstractly defined properties of a group are known.

The *generic group model* considers a class of algorithms operating on an algebraic group by performing only certain operations, such as applying the group law and testing for equality of group elements. In particular, generic algorithms do not exploit specific properties of a particular representation of group elements. This is modeled by treating the group as a black box. The algorithm interacts with this black box in order to perform computations. Since all computations are independent of a particular representation of group elements, an algorithm solving a problem in the generic group model is capable of solving the problem in any concrete instantiation of the group, thus is *generic*. Well-known examples for such algorithms are Pollard's Rho [Pol75] or the Baby-Step Giant-Step [Sha71] algorithm for computing discrete logarithms.

Generic groups and their extensions to generic bilinear groups and generic rings are used as tools to analyze the validity of classical and newly introduced computational hardness assumptions.

There are basically two motivations for the study of cryptographic assumptions in these black-box models.

1. THE CRYPTOGRAPHER'S MOTIVATION. A proof that a computational problem cannot be solved by a reasonably restricted class of algorithms may be
   seen as support towards the assumption that the problem is also hard to solve
   in the standard model (where algorithms are modeled as Turing machines).
   Thus, a proof in an idealized model may strenghten the trust in the validity
   of a hardness assumption.

2. THE CRYPTANALYST'S MOTIVATION. Showing that a certain class of algorithms is not able to solve a certain problem efficiently is a helpful insight
   for the search for cryptanalytic algorithms. Proving that a certain class of
   algorithms is not capable of solving a problem efficiently can for instance
   help to falsify approaches for finding efficient algorithms.

Several fundamental questions concerning generic models of computation arise
with these motivations, which are addressed in this book.

- Are the generic group model and the generic ring model, as currently used
  in the cryptographic literature, a reasonable abstraction of reality? What are
  the limitations of these models, and to which extent can results be adopted
  to the "real world"?

- Can we relax these models to bring them closer to the standard model? The
  challenge here is to make the models as realistic as possible, while still being
  able to prove meaningful results that we are currently not able to obtain in
  the standard model. How close can we get to reality?

- One of the most important open questions in cryptography is whether solving the computational Diffie-Hellman problem is equivalent to computing
  discrete logarithms. A *generic* reduction from the discrete logarithm problem to the Diffie-Hellman problem would be particularly interesting, since
  it would imply the equivalence of both problems in *all* groups. Does such a
  reduction exist?

1.1 Outline of this Book

In Chapter 2 we start with describing a general black-box model of computation,
which provides a unified description for various models, and allows us to instantiate all generic group and generic ring models that are currently used in the literature. The general model serves also to provide a unified notation for the different
models presented and used in this book.

In Chapter 3 the *black-box ring extraction problem* is considered. This problem is interesting, since an efficient algorithm for this problem would imply the equivalence of the discrete logarithm and the Diffie-Hellman problem in all finite cyclic groups, and the inexistence of ring-homomorphic one-way permutations. We show that an efficient algorithm for the black-box ring extraction problem over $\mathbb{Z}_N$ yields an efficient algorithm finding a factor of N. This shows that there is no efficient reduction from the discrete logarithm to the Diffie-Hellman problem that works in any group, unless factoring integers is easy. This chapter is based on a work presented at ICALP 2008 [AJR08]. In contrast to the original paper, we give a new reduction which is simpler, more general, and tighter than the original one.

Another fundamental open problem in cryptology is whether solving the RSA problem is equivalent to factoring. It was shown that indeed this equivalence holds in the *generic ring model* [LR06, AM09]. In Chapter 4 we consider the question whether these results suggest that breaking RSA is also equivalent to factoring in the standard model. Our answer is negative, since we show that there exist a simple and *natural* (i.e., not contrived) computational problem which is easy to solve in the standard model, but equivalent to factoring in the generic ring model. Concretely, we show that computing the *Jacobi symbol* is equivalent to factoring with respect to generic ring algorithms. We achieve this result by proving a general theorem which states that a large class of decisional problems is hard in the generic ring model. As a further application of this theorem, we show that no generic ring algorithm can solve the *quadratic residuosity* problem without (essentially) factoring the modulus. This chapter is based on [JS09], but we give alternate (and significantly simpler) proofs for the two key lemmas leading to the main result.

Chapter 5 considers a generalized version of the *decisional composite residuosity* problem, which captures several variants that were proposed in the literature. These problems have recently found several interesting cryptographic applications. Due to a different algebraic setting, the main theorem from Chapter 4 can not be applied. Therefore we have to devise a different technique to relate the generic decisional composite residuosity problem to the so-called Hensel-RSA problem.

A caveat of existing generic group models is that not all currently known algorithms for solving computational problems are captured. Thus, hardness results are of very limited significance. In Chapter 6 we extend the notion of semi-generic groups introduced in [JR10]. We propose novel idealizations of groups and bilinear groups, which capture *all currently known algorithms*, and thus allow to derive more meaningful results than the classical models. We show the applicability of the new models by analyzing exemplarily several relevant computational problems, and show for instance that solving the computational Diffie-Hellman problem semi-generically is equivalent to solving the discrete logarithm problem.

2 Black-Box Models of Computation

Several abstract models of computation have been proposed in the literature to capture the notion of "generic groups" or their extensions [BS84, Nec94, Sho97, Mau05]. The black-box model we are going to describe in the sequel is a generalization of the model introduced by Maurer in [Mau05], which in turn can be seen as a generalization of the model of [Sho97]. While Maurer's model captures only the case where a generic algorithm operates on elements of a *single* algebraic structure, like a group or a ring, we extend this model in order to be able to describe settings where more than one algebraic structure is considered, as in pairing-based cryptography [DBS04], for instance.

2.1 A General Black-Box Model

Our black-box model is characterized by $q + 2$ sets

$$S_1, \ldots, S_q, \Pi, \Sigma.$$

Each set S_i contains elements of an algebraic structure, like a group or a ring. The set Π contains functions representing computations that can be performed on elements of $\bigcup_{i=1}^{q} S_i$, and Σ contains relations on elements of $\bigcup_{i=1}^{q} S_i$.

An algorithm $\mathscr{A}$ intending to solve a computational problem over $S_1, \ldots, S_q$ interacts with an oracle $\mathscr{O}$. The oracle keeps internal lists storing elements of $\bigcup_{i=1}^{q} S_i$. It may be queried by $\mathscr{A}$ to compute functions from Π and relations from Σ on elements of $\bigcup_{i=1}^{q} S_i$.

2.1.1 Internal State of $\mathscr{O}$

For each set S_i, the oracle maintains a list L_i storing elements of S_i. We write $L_{i,j}$ to denote the j-th entry of list L_i.

The oracle receives as input q vectors $\vec{x}_1, \ldots, \vec{x}_q$. Each vector

$$\vec{x}_i = (x_{i,1}, \ldots, x_{i,k_i}) \in S_i^{k_i}$$

consists of k_i elements of S_i. The lists are initialized such that

$$L_{i,j} = x_{i,j} \quad \text{for } i \in \{1, \ldots, q\} \text{ and } j \in \{1, \ldots, k_i\}.$$

The distribution of each vector $\vec{x}_i$ depends on the considered computational problem, and will be described seperately for each problem considered in the sequel.

2.1.2 Querying Operations and Relations

The algorithm may query the oracle to perform computations and relations on elements stored in variables. In order to query a computation, the algorithm submits $2n$ indices $(i_1, j_1, \ldots, i_n, j_n)$ and some n-ary function

$$f : S_{i_1} \times \cdots \times S_{i_n} \to S_i$$

with $f \in \Pi$. If $f(L_{i_1, j_1}, \ldots, L_{i_n, j_n})$ is undefined[1] then the oracle returns an error symbol $\bot$. Otherwise it computes

$$f(L_{i_1, j_1}, \ldots, L_{i_n, j_n}).$$

and appends the result to list i.

The algorithm may also issue relation queries from set Σ. To this end, it submits $2n$ indices $(i_1, j_1, \ldots, i_n, j_n)$ and an n-ary relation

$$\rho : S_{i_1} \times \cdots \times S_{i_n} \to \{0, 1\}$$

with $\rho \in \Sigma$. The oracle returns $\rho(L_{i_1, j_1}, \ldots, L_{i_n, j_n})$.

2.1.3 Notation

We write

$$\mathcal{O}(S_1, \ldots, S_q, \Pi, \Sigma; \vec{x}_1, \ldots, \vec{x}_q)$$

to denote that we consider an oracle storing elements of $S_1, \ldots, S_q$ in the lists $L_1, \ldots, L_q$, providing operations from Π and relations from Σ, and whose lists are initialized with $(\vec{x}_1, \ldots, \vec{x}_q)$.

When it is clear from the context, then we omit $S_1, \ldots, S_q$ and (Π, Σ), and write $\mathcal{O}(\vec{x}_1, \ldots, \vec{x}_q)$ or simply $\mathcal{O}$ for short. In scenarios where $q = 1$, that is, when only a single algebraic structure is considered, then we omit the subscripts indicating different sets S_i and write S, L, and $\vec{x} = (x_1, \ldots, x_k)$.

We write $\mathcal{A}^{\mathcal{O}}(X)$ to denote that an algorithm $\mathcal{A}$ has oracle access to $\mathcal{O}$, and receives X as additional input (X may be the order of the considered group, for instance).

If A is a set, then we write $a \xleftarrow{\$} A$ to denote that a is chosen uniformly random from A.

---

[1] For instance, S_i may be a semi-group and f tries to compute the inverse of a non-invertible element.

2.2 The Generic Group Model

Let $\mathbb{G} = (S, \circ)$ be a group. In the *generic group model* (GGM) an algorithm interacts with an oracle

$$\mathcal{O}(S, \Pi, \Sigma; \vec{x}),$$

where $\Pi = \{\circ\}$ and $\Sigma = \{=\}$. Here $=$ denotes the (binary) equality relation. The vector $\vec{x}$ is chosen according to the considered computational problem, and contains usually at least a generator of the group $(S, \circ)$.

2.2.1 The Generic Hardness of Computing Discrete Logarithms

In this section, we will present a classical result from [Nec94, Sho97, Mau05] on the hardness of computing discrete logarithms with generic group algorithms. This should serve as an illustration how proofs in the generic group model usually work, since it uses the standard "polynomial degree" argument which has been used for several further results in the generic model, in particular for the hardness conditions of [RLB$^+$08] which capture nearly all cryptographically relevant problems over groups.

In the sequel let us consider the discrete logarithm problem in a finite cyclic group $\mathbb{G} = (S, \circ)$ of order N. Let g be a generator of $\mathbb{G}$. The discrete logarithm problem is: given $g, g^x \in S$, where $x \xleftarrow{\$} \mathbb{Z}_N$ is uniformly random, determine $x = \log_g g^x$.

To analyze the discrete logarithm problem in the generic group model, we consider algorithms interacting with a generic group oracle

$$\mathcal{O}(S, \Pi, \Sigma; \vec{x}),$$

where $\vec{x} = (g, g^x)$ with $x \xleftarrow{\$} \mathbb{Z}_N$, $\Pi = \{\circ\}$ and $\Sigma = \{=\}$.

Definition 2.1

We say that the generic discrete logarithm problem is (ε_{dl}, t)-hard, if

$$\Pr[\mathscr{A}^{\mathcal{O}}(N) = x] \leq \varepsilon_{dl}$$

for all algorithms $\mathscr{A}$ running in time t.

Theorem 2.1

Let $\mathbb{G} = (S, \circ)$ be a cyclic group of order N, and let P be the largest prime factor of N. Then solving the discrete logarithm problem generically is (ε_{dl}, t)-hard with

$$\varepsilon_{dl} \leq 1/N + \frac{t^2 + 3t + 2}{P}.$$

Proof outline. We proceed in a sequence of games (cf. [Sho04, BR06]). We start with the original discrete logarithm security experiment. We end up in a game where the oracle performs all computation independent of the discrete logarithm challenge $x \in \mathbb{Z}_N$, thus any algorithm can only guess the discrete logarithm x. We conclude by showing that each game is indistinguishable from the previous one, except with probability $(t^2 + 3t + 2)/P$, which yields the result.

PROOF. In the following let $\mathcal{O}_i$ denote the oracle that $\mathcal{A}$ interacts with in Game i.

Game 0. This is the original discrete logarithm security experiment. That is, $\mathcal{O}_0$ proceeds exactly like Oracle $\mathcal{O}$ from Definition 2.1. Thus we have

$$\Pr[\mathcal{A}^{\mathcal{O}_0}(N) = x] = \varepsilon$$

for some ε.

Game 1. In this game we replace Oracle $\mathcal{O}_0$ with

$$\mathcal{O}_1(\mathbb{Z}_N, \Pi, \Sigma; \vec{x}),$$

where $\vec{x} = (1, x)$ with $x \xleftarrow{\$} \mathbb{Z}_N$, $\Pi = \{+\}$ and $\Sigma = \{=\}$. Thus, $\mathcal{O}_1$ performs all computations in the group $(\mathbb{Z}_N, +)$ using generator $1 \in \mathbb{Z}_N$. Since the groups $(\mathbb{Z}_N, +)$ and $\mathbb{G}$ are isomorphic, we have

$$\Pr[\mathcal{A}^{\mathcal{O}_1}(N) = x] = \Pr[\mathcal{A}^{\mathcal{O}_0}(N) = x]$$

Game 2. We replace Oracle $\mathcal{O}_1$ with $\mathcal{O}_2$. Instead of x, Oracle $\mathcal{O}_2$ uses variable X as a wildcard character. It simulates $\mathcal{O}_1$ as follows.

- The list L is initialized with $\vec{x} = (1, X)$.

- Whenever the algorithm queries to apply the group law to two list elements L_i, L_j, the oracle computes

$$L_i + L_j \in \mathbb{Z}_N[X].$$

  Note that we can write each list element L_i as a polynomial $L_i(X) = a_i X + b_i \in \mathbb{Z}_N[X]$ of degree 1, with $a_i, b_i \in \mathbb{Z}_N$,

- Whenever the algorithm issues an equality test query of two list elements $L_i(X), L_j(X) \in \mathbb{Z}_N[X]$, the oracle returns 1 if

$$L_i(x) = L_j(x),$$

  and 0 otherwise.

For algorithm $\mathscr{A}$ oracle $\mathcal{O}_2$ is perfectly indistinguishable from oracle $\mathcal{O}_1$. This implies that we have

$$\Pr[\mathscr{A}^{\mathcal{O}_2}(N) = x] = \Pr[\mathscr{A}^{\mathcal{O}_1}(N) = x].$$

Game 3. We replace Oracle $\mathcal{O}_2$ with a simulator $\mathcal{O}_3$ which performs all computations independent of x. $\mathcal{O}_2$ proceeds exactly like $\mathcal{O}_2$, except for the following.

- Whenever the algorithm issues an equality test query of two list elements $L_i(X), L_j(X) \in \mathbb{Z}_N[X]$, the oracle returns 1 if

$$L_i(X) = L_j(X),$$

where equality is tested by checking whether $(a_i, b_i) = (a_j, b_j)$, and 0 otherwise.

Note that the simulator $\mathcal{O}_3$ simulates $\mathcal{O}_2$ perfectly, unless the algorithm issues an equality test query for two list elements $L_i(X) = a_i X + b_i$ and $L_j(X) = a_j X + b_j$ such that $L_i(X) \neq L_j(X)$ and $L_i(x) \equiv L_j(x) \bmod N$, or equivalently

$$(L_i - L_j)(X) \not\equiv 0 \bmod N \qquad \text{but} \qquad (L_i - L_j)(x) \equiv 0 \bmod N.$$

We denote this event with $\mathscr{F}$. Note that we have

$$\begin{aligned}
\Pr[\mathscr{A}^{\mathcal{O}_3}(N) = x] &= \Pr[\mathscr{A}^{\mathcal{O}_3}(N) = x \wedge \neg\mathscr{F}] + \Pr[\mathscr{A}^{\mathcal{O}_3}(N) = x \wedge \mathscr{F}] \\
&\leq \Pr[\mathscr{A}^{\mathcal{O}_3}(N) = x \wedge \neg\mathscr{F}] + \Pr[\mathscr{F}].
\end{aligned}$$

Note that by applying only addition and subtraction operations, the algorithm computes polynomials $L_i(X)$ of degree at most one. Thus, for each (i, j) the polynomial $\Delta_{i,j}(X) = (L_i - L_j)(X)$ has at most one root modulo P, where P is the largest prime factor of N. $\mathscr{A}$ issues at most t oracle queries, thus there are at most $t + 2$ polynomials in the list L. Therefore there are at most $(t + 2)(t + 1) = t^2 + 3t + 2$ possible difference polynomials $\Delta_{i,j}$. Since x is uniformly random and independent of $\mathscr{A}$'s view (and thus independent of all (a_i, b_i)), we have

$$\Pr[\mathscr{F}] \leq \frac{t^2 + 3t + 2}{P}.$$

Moreover, all computations of $\mathscr{A}$ are independent of x, thus we have

$$\Pr[\mathscr{A}^{\mathcal{O}_3}(N) = x \wedge \neg\mathscr{F}] = 1/N.$$

Summing up, we obtain that the success probability of $\mathscr{A}$ when interacting with oracle $\mathscr{O}_3$ is bounded by

$$\Pr[\mathscr{A}^{\mathscr{O}_3}(N) = x] \leq 1/N + \frac{t^2 + 3t + 2}{P}$$

which implies

$$\varepsilon \leq 1/N + \frac{t^2 + 3t + 2}{P}.$$

$\square$

2.2.2 Generic Group Models in the Literature

Several formalizations of the notion of generic groups have been proposed in the literature [Nec94, Sho97, Mau05]. The model of Nechaev [Nec94] is focused only on the analysis of the discrete logarithm problem. Shoup [Sho97] has introduced a more general model, where group elements are encoded by random bit strings, which models the space requirements of certain algorithms, such as Pollard's Rho [Pol75], more accurately. The fact that elements are encoded by *random* strings has been criticised [Fis00], since it may be abused in security proofs, for instance to implement a *random oracle* [BR93]. The most flexible description of the model was formulated by Maurer [Mau05], this model is equivalent to the one described above. One can however show that the model of Maurer (and thus also the above model) and the model of Shoup are equivalent as long as

1. only the algebraic properties are relevant, but not the encoding of group elements as random strings from [Sho97], and

2. only the total running time of algorithms is analyzed, but not the space requirements.

See [JS08] for details.

2.3 The Generic Ring Model

Let $R = (S, +, \cdot)$ be a ring. In the *generic ring model* (GRM) an algorithm interacts with an oracle

$$\mathscr{O}(S, \Pi, \Sigma; \vec{x}),$$

where $\Pi = \{+,-,\cdot,\div\}$ and $\Sigma = \{=\}$. Here $-$ is the addition with additive inverses, $\div$ is the multiplication with multiplicative inverses, and $=$ denotes the equality relation.

2.3.1 Comparing the Generic Ring Model to the Generic Group Model

The generic ring model (GRM) is an extension of the generic *group* model (GGM). Despite many similarities to the GGM, showing the hardness of computational problems in the GRM seems to be more involved than standard proofs in the GGM. The reason is that a typical proof in the GGM (cf. Section 2.2.1 and [Sho97, Mau05, RLB+08], for instance) introduces a simulation game where group elements are replaced with polynomials (or sometimes rational functions, given as straight line programs) that are (implicitly) evaluated with some group elements corresponding to a given problem instance. A key argument in these proofs is that, by construction of the simulator, the degree of these polynomials cannot exceed a certain small bound (often degree one or two). Following Shoup's seminal work [Sho97], a lower bound on the success probability of any generic group algorithm for the given problem is then derived by bounding the number of roots of these polynomials. Usually the bound is useful only if the number of roots is sufficiently small.

For instance, in the proof of Theorem 2.1 in Section 2.2.1, a key argument was that by performing a sequence of addition operations on the initial list elements 1 and $X \in \mathbb{Z}_N[X]$ the algorithm is only able to compute polynomials of the form $aX + b$ having degree one. This technique is the basis for many proofs in the generic group model: the number of roots of polynomials is kept small by performing only addition operations on polynomials of bounded degree in the simulation game. Sometimes also a small bounded number of multiplications is allowed, for instance when the model is extended to groups with bilinear pairing map, as done in [KSW08, RLB+08, Boy08, BB08] (see also Section 2.4).

However, in the generic ring model we do not want to bound the number of allowed multiplications. Thus, by repeated squaring an algorithm may compute polynomials of exponential degree. In this case we cannot obtain a useful bound by counting the number of roots.

2.3.2 Generic Ring Models in the Literature

There are several variants of the GRM in the literature. The model described above is equivalent to the models used in [AM09] and [JS09]. This is the strongest and most general variant, in the sense that it gives the algorithm more power than other models.

In [LR06] a slightly weaker model was considered, where $\Pi = \{+,-,\cdot\}$ and $\Sigma = \{=\}$, thus the algorithm may not multiply by multiplicative inverses of ring elements. In [Bro05] a model is considered where $\Pi = \{+,-,\cdot,\div\}$ and $\Sigma = \emptyset$, thus no equality tests are allowed. Damgård and Koprowski [DK02] considered only the multiplicative group, that is, a model where $\Pi = \{\cdot,\div\}$ and $\Sigma = \{=\}$.

2.4 The Generic Bilinear Group Model

Let $\mathbb{G}_1 = (S_1, \circ_1)$, $\mathbb{G}_2 = (S_2, \circ_2)$, $\mathbb{G}_3 = (S_3, \circ_3)$ be groups.

Definition 2.2

A *pairing* is a map $e : \mathbb{G}_1 \times \mathbb{G}_2 \to \mathbb{G}_3$ with the following properties:

1. *Bilinearity*: $\forall (a,b) \in \mathbb{G}_1 \times \mathbb{G}_2$ and $x_1, x_2 \in \mathbb{Z}_p$ holds that $e(a^{x_1}, b^{x_2}) = e(a,b)^{x_1 x_2}$.

2. *Non-degeneracy*: $g_3 := e(g_1, g_2)$ is a generator of $\mathbb{G}_3$, if g_1 and g_2 are generators.

3. e is efficiently computable.

Following [GPS08], we distinguish three different types of bilinear group settings:

- Type 1: $\mathbb{G}_1 = \mathbb{G}_2$. We will call this the setting with *symmetric* bilinear map.
- Type 2: $\mathbb{G}_1 \neq \mathbb{G}_2$ and there is an *efficiently computable* isomorphism $\psi : \mathbb{G}_1 \to \mathbb{G}_2$.
- Type 3: $\mathbb{G}_1 \neq \mathbb{G}_2$ and there is no efficiently computable isomorphism $\psi : \mathbb{G}_1 \to \mathbb{G}_2$.

The *generic bilinear group model* is defined as follows.

Type-I Settings. When considering Type I settings (i.e., symmetric bilinear groups), then an algorithm interacts with an oracle

$$\mathcal{O}((S_1, S_3), \Pi, \Sigma; \vec{x}_1, \vec{x}_3),$$

where $\Sigma = \{=\}$ contains the equality relation, $\Pi = \{\circ_1, \circ_3, e\}$, and e is a symmetric bilinear map $e : S_1 \times S_1 \to S_3$.

Type-II Settings. When considering Type II settings (asymmetric bilinear groups with efficient isomorphism $\psi : S_1 \to S_2$), then an algorithm interacts with an oracle

$$\mathcal{O}((S_1, S_2, S_3), \Pi, \Sigma; \vec{x}_1, \vec{x}_2, \vec{x}_3),$$

where $\Pi = \{\circ_1, \circ_2, \circ_3, e, \psi\}$, $\Sigma = \{=\}$ contains the equality relation, e is a bilinear map $e : S_1 \times S_2 \to S_3$, and ψ is the isomorphism $\psi : \mathbb{G}_1 \to \mathbb{G}_2$.

Type-III Settings. The model for Type III settings is identical to the Type II-model, except that $\Pi = \{\circ_1, \circ_2, \circ_3, e\}$, thus there is no efficiently computable isomorphism ψ.

2.4.1 Generic Bilinear Group Models in the Literature

Models similar to the above have been used in [KSW08, RLB$^+$08, Boy08, BB08] to analyze novel cryptographic assumptions from pairing-based cryptography. All these models are a straightforward extensions of the generic group model.

3 On Black-Box Ring Extraction and Integer Factorization

The black-box ring extraction (BBRE) problem is the problem of extracting a secret ring element from a black-box which may be queried to perform the ring operations and equality tests on internally stored ring elements. This problem has at least two important applications to cryptography (see Section 3.1 for details).

1. An efficient algorithm for the black-box ring extraction problem implies the equivalence of the discrete logarithm and the Diffie-Hellman problem.

2. At the same time this implies the inexistence of ring-homomorphic one-way permutations.

The BBRE problem has been studied in a long line of research [Mau94, BL96, MW98, MW99, MR07, AJR08], which we describe in more detail in Section 3.3. It is, however, still unknown whether the known subexponential-time algorithms for BBRE are optimal. In this chapter we describe an efficient reduction from the problem of factoring integers to the black-box ring extraction problem. This result has the following consequences.

1. It implies that there is no generic reduction from the discrete logarithm to the Diffie-Hellman problem, unless factoring integers is easy.

2. It may be seen as an indicator that ring-homomorphic one-way permutations may exist.

The existence of such a reduction has been conjectured by Stefan Wolf in his PhD thesis [Wol99, Conjecture 10.1]. A preliminary version of the results described in this chapter was presented at ICALP 2008 [AJR08]. Compared to the proof [AJR08], we give an improved reduction which is more general, tighter, and much simpler.

3.1 Motivation

A famous open question in cryptology is whether breaking the Diffie-Hellman protocol [DH76] is as hard as computing discrete logarithms. The discrete logarithm problem can be reduced to the Diffie-Hellman problem by assuming a *Diffie-Hellman oracle* DH solving the Diffie-Hellman problem, and showing that there exists an algorithm solving the discrete logarithm problem efficiently using DH. If the reduction algorithm is *generic*, that is, does not exploit specific properties of a given representation of group elements, then such an algorithm would imply that breaking the Diffie-Hellman protocol is equivalent to computing discrete logarithms in *any* group. Maurer and Wolf [MW99] have shown that any such reduction has complexity $\Omega(\sqrt{p})$, if the group order is divisible by p^2 for a large prime p. However, for all other groups,in particular many cryptographically relevant groups whose order is not divisible by a large square, it is unknown whether an efficient reduction exists.

3.1.1 Is Solving the Diffie-Hellman Problem Equivalent to Computing Discrete Logarithms?

Let $\mathbb{G} = (S, \circ)$ be a finite cyclic group of order $N \in \mathbb{N}$ with generator g, and let ϕ be the isomorphism

$$\phi : (\mathbb{Z}_N, +) \to \mathbb{G}, \quad a \mapsto g^a.$$

In this notation, applying the group law to two elements $\phi(a)$ and $\phi(b)$ yields $\phi(a+b)$, and the discrete logarithm problem in $\mathbb{G}$ is to compute $a \in \mathbb{Z}_N$ on input $\phi(1)$ and $\phi(a)$.

Now let us assume a Diffie-Hellman oracle DH for $\mathbb{G}$ taking as input a pair $(\phi(a), \phi(b))$, and returning $\phi(ab)$. Then the group $\mathbb{G}$ together with the Diffie-Hellman oracle is isomorphic to the *ring* $\mathbb{Z}_N$, where addition in $\mathbb{Z}_N$ can be performed by applying the group law

$$\phi(a+b) = \phi(a) \circ \phi(b)$$

and the multiplication operation can be performed by querying the Diffie-Hellman oracle

$$\phi(ab) = \mathrm{DH}(\phi(a), \phi(b)).$$

Thus we see that the question whether there exists a *generic* reduction from the discrete logarithm problem to the Diffie-Hellman problem corresponds to the question whether there exists an algorithm inverting any function ϕ by exploiting only the *structure* of the ring $(S, \circ, \mathrm{DH}) \cong (\mathbb{Z}_N, +, \cdot)$.

3.1.2 Do Ring-Homomorphic One-Way Permutations Exist?

Now let us consider a one-way permutation $\phi : S \to S$. We say that ϕ is *homomorphic*, if

1. S exhibits an algebraic structure (e.g., S is a group or a ring, for instance),

2. the algebraic operations on S can be computed efficiently, and

3. ϕ preserves the algebraic structure of S, that is, we have

$$\phi(a) \circ \phi(b) = \phi(a \circ b)$$

for all $a, b \in S$ and an algebraic operation $\circ$ provided by S

Under various complexity assumptions, there are several *group*-homomorphic one-way permutations, which are homomorphic with respect to *one* algebraic operation over S. Well-known examples are the RSA permutation [RSA78] over $\mathbb{Z}_N^*$ or the Rabin permutation [Rab79] over the quadratic residues modulo N for suitable $N \in \mathbb{N}$, for instance. However, it is a long-standing open question whether there exist *ring*-homomorphic one-way permutations, which are homomorphic with respect to two operations such that the resulting algebraic structure forms a ring. An important step towards an answer of this question is to answer the following questions.

- Is the *structure* of a ring sufficient to invert any ring-homomorphic permutation ϕ efficiently? This would imply the inexistence of such one-way permutations.

- Which conditions must be satisfied by a ring such that there is no efficient inversion algorithm?

Motivated by the above observations, we define the *black-box ring extraction problem* as the problem of inverting a ring-homomorphism ϕ by exploiting solely the structure of the given ring, without exploiting specific properties of a representation of ring elements.

3.2 Results of This Chapter

In preliminary work [AJR08] it was shown that there is no algorithm solving the BBRE problem for rings of characteristic N without (essentially) revealing a factor of N, thus in this sense the $\mathbb{Z}_N$-variant of the algorithm described in [BL96], which requires to factorize N, is optimal. Hence, the black-box extraction problem can not be solved efficiently in general, unless there is an efficient algorithm for integer factorization. We note that in the model of [AJR08] an explicit operation for computing multiplicative inverses was excluded. Since inverses can be computed efficiently in many rings (such as $\mathbb{Z}_n$), it is desirable to include this operation.

In this chapter, we devise a novel technique to reduce the problem of factoring an integer N to the black-box ring extraction problem in the ring $\mathbb{Z}_N$. We obtain a tighter reduction than [AJR08], by a much simpler argument, and in addition allow the computation of multiplicative inverses. In combination with [AJR08], the results in this chapter can be generalized to arbitrary rings of characteristic N.

3.3 Related Work

Nechaev [Nec94] and Shoup [Sho97] studied the black-box extraction problem for *groups*, showing that any generic algorithm has to perform $\Omega(\sqrt{p})$ group operations to solve this problem, where p is the largest prime factor of the order of the group (see also Section 2.2.1). This bound essentially matches the running time of well-known generic algorithms for the discrete logarithm problem [Sha71, Pol75]. Hence, the structure of a group is in general not sufficient to solve the black-box extraction problem efficiently. Efficient reductions are known only if the group order meets certain properties [den90, MW96], or if a certain auxiliary information depending on the group order is given [Mau94].

Boneh and Lipton [BL96] applied a technique due to Maurer [Mau94] to describe an algorithm solving the black-box extraction problem over *prime fields* $\mathbb{F}_p$ in subexponential-time in $\log p$ under a (plausible) number-theoretic conjecture on the distribution of smooth integers. Hence, in comparison to the results of Nechaev [Nec94] and Shoup [Sho97], the additional structure of a field helps to solve the problem considerably more efficiently than in the case where the underlying algebraic structure is a group. It is unknown whether there are more efficient algorithms than the one proposed in [BL96]. Maurer and Raub [MR07] augmented the work of [BL96] from prime fields to extension fields, by reducing the BBRE problem over an extension field to the BBRE problem over the underlying prime field.

As mentioned by Boneh and Lipton, the algorithm described in [BL96] can also be extended from prime fields to rings $\mathbb{Z}_N$ if N is a *squarefree* composite integer, by first factoring $N = \prod_{i=1}^{k} P_i$ (which can be done in subexponential time with current factoring algorithms) and then solving the BBRE problem for all fields $\mathbb{Z}_{P_i}$, $i \in \{1, \ldots, k\}$. The solution for $\mathbb{Z}_N$ can then be obtained by Chinese remaindering.

Maurer and Wolf [MW99] considered the BBRE problem for $\mathbb{Z}_N$ when showing that there is no efficient generic reduction from the discrete logarithm to the Diffie-Hellman problem if the group order is divisible by a large multiple prime factor.

3.4 The Black-Box Ring Extraction Problem

We formalize the BBRE problem over a ring $(R, +, \cdot)$ using the framework from Section 2.1. Consider a game between an algorithm $\mathscr{A}$ for the BBRE problem and an oracle

$$\mathscr{O}(R, \Pi, \Sigma; (1, x))$$

where Π contains the operations addition (with inverses) and multiplication (with inverses) in R, denoted with $\Pi = \{+, -, \cdot, \div\}$, and Σ provides the equality relation in R, i.e. $\Sigma = \{=\}$. The element $1 \in R$ given as input to $\mathscr{O}$ is a generator of $(R, +)$ and enables the algorithm to compute any element of R inside $\mathscr{O}$. The value $x \overset{\$}{\leftarrow} R$ is chosen uniformly random from R.

The black-box ring extraction problem is to extract x from $\mathscr{O}(R, \Pi, \Sigma; (1, x))$.

Definition 3.1
We say that algorithm $\mathscr{A}$ (ε, t)-solves the black-box ring extraction problem for R, if $\mathscr{A}$ makes at most t oracle queries, and

$$\Pr\left[\mathscr{A}^{\mathscr{O}(R, \Pi, \Sigma; (1, x))}(N) = x\right] \geq \varepsilon.$$

3.5 Main Result

Now let us consider the BBRE problem for the ring $R = \mathbb{Z}_N$. As explained before, this case has the most interesting cryptographic applications.

Theorem 3.1
Let $N \in \mathbb{N}$ be a composite integer, and let P be the smallest prime factor of N. Suppose there exists an algorithm $\mathscr{A}$ (ε, t)-solving the black-box ring extraction problem over $\mathbb{Z}_N$. Then there exists a factoring algorithm $\mathscr{B}$, which returns P on input N with probability

$$\Pr[\mathscr{B}(N) = P] \geq \varepsilon - P/N$$

by performing t ring operations and equality tests in $\mathbb{Z}_N$, and t^2 computations of a greatest common divisor on $\log_2 N$-bit integers.

Proof outline. We prove the theorem by a short *sequence of games* [Sho04, BR06]. Game 0 corresponds to the original BBRE game, where the algorithm $\mathscr{A}$ interacts with an oracle $\mathscr{O}(\mathbb{Z}_N, \Pi, \Sigma; (1, x))$. In Game 1 we add an event abort and terminate the game, if abort occurs. Then we describe Game 2, which is perfectly indistinguishable from Game 1. In Game 2 the algorithm interacts with an oracle

$$\mathcal{O}'(\mathbb{Z}_P, \Pi, \Sigma, (1, x \bmod P)),$$

which performs all computations in the ring $\mathbb{Z}_P$, and thus reveals only $x \bmod P$ (information-theoretically). Therefore $\mathcal{A}$ does not learn anything about $x \bmod N/P$, which implies that the probability that $\mathcal{A}$ outputs x correctly in Game 0 is at most $\varepsilon \leq P/N + \Pr[\text{abort}]$. Finally we show that there exists an algorithm $\mathcal{B}$ running $\mathcal{A}$ as a subroutine and finding a factor of N with probability $\Pr[\text{abort}]$ whose running time is bounded by a polynomial in t.

PROOF.

Game 0. This is the original BBRE experiment played between algorithm $\mathcal{A}$ and Oracle $\mathcal{O}$. By assumption, we have

$$\Pr\left[\mathcal{A}^{\mathcal{O}(\mathbb{Z}_N, \Pi, \Sigma; (1,x))}(N) = x\right] \geq \varepsilon.$$

Game 1. We replace oracle $\mathcal{O}(\mathbb{Z}_N, \Pi, \Sigma; (1,x))$ with $\mathcal{O}_1(\mathbb{Z}_N, \Pi, \Sigma; (1,x))$. $\mathcal{O}_1$ proceeds exactly like $\mathcal{O}$, except for the following. Each time a ring element $y \in \mathbb{Z}_N$ is appended to list L_1, oracle $\mathcal{O}_1$ tests whether there exists some ring element $y' \in L_1 \cup \{0\}$ such that

$$y \not\equiv y' \bmod N \qquad \text{and} \qquad y \equiv y' \bmod P. \tag{3.1}$$

(We may assume that oracle $\mathcal{O}_1$ is computationally unbounded, and thus can compute P on input N, since this oracle is never implemented and thus need not be efficient). If Condition (3.1) holds, then $\mathcal{O}_1$ raises event abort and terminates.

Since Game 1 and Game 0 proceed identical until abort is raised, we have

$$\left|\Pr\left[\mathcal{A}^{\mathcal{O}(\mathbb{Z}_N, \Pi, \Sigma; (1,x))}(N) = x\right] - \Pr\left[\mathcal{A}^{\mathcal{O}_1(\mathbb{Z}_N, \Pi, \Sigma; (1,x))}(N) = x\right]\right| \leq \Pr[\text{abort}]$$

Note that in Game 1 it holds that

$$y \equiv y' \bmod N \iff y \equiv y' \bmod P \quad \text{and} \quad y \equiv 0 \bmod N \iff y \equiv 0 \bmod P$$

for all elements $y, y' \in L_1$, as the game is aborted otherwise.

Game 2. In this game, we replace $\mathcal{O}_1(\mathbb{Z}_N, \Pi, \Sigma; (1,x))$ with oracle

$$\mathcal{O}_2(\mathbb{Z}_P, \Pi, \Sigma, (1, x \bmod P)).$$

$\mathcal{O}_2$ proceeds exactly like $\mathcal{O}_1$, except that it represents elements of $\mathbb{Z}_N$ by elements of $\mathbb{Z}_P$, and performs all computations modulo P.

Recall that, due to the modifications introduced in Game 1, we have

$$y \equiv y' \bmod N \iff y \equiv y' \bmod P \quad \text{and} \quad y \equiv 0 \bmod N \iff y \equiv 0 \bmod P.$$

Therefore $\mathscr{O}_2$ responds to all queries issued by $\mathscr{A}$ exactly like $\mathscr{O}_1$ does. Thus $\mathscr{O}_2$ simulates $\mathscr{O}_1$ perfectly. This implies

$$\Pr\left[\mathscr{A}^{\mathscr{O}_1(\mathbb{Z}_N,\Pi,\Sigma;(1,x))}(N) = x\right] = \Pr\left[\mathscr{A}^{\mathscr{O}_2(\mathbb{Z}_P,\Pi,\Sigma,(1,x \bmod P))}(N) = x\right].$$

In Game 2 algorithm $\mathscr{A}$ can (information-theoretically) learn only $x \bmod P$. Since x is distributed uniformly over $\mathbb{Z}_N$ and P divides N, this implies that

$$\Pr\left[\mathscr{A}^{\mathscr{O}_2(\mathbb{Z}_P,\Pi,\Sigma,(1,x \bmod P))}(N) = x\right] \le P/N.$$

Summing up probabilities from Game 0 to Game 2, we obtain that

$$\Pr[\text{abort}] \ge |\varepsilon - P/N|.$$

It remains to show that there exists a factoring algorithm $\mathscr{B}$ returning P on input N with probability $\Pr[\text{abort}]$. $\mathscr{B}$ proceeds as follows. It receives as input N and samples $x \xleftarrow{\$} \{0, \dots, N-1\}$ uniformly random. Then it runs $\mathscr{A}$ as a subroutine, by implementing the generic ring oracle from Game 0 for $\mathscr{A}$ (representing elements of $\mathbb{Z}_N$ by integers from the set $\{0, \dots, N-1\}$).

Each time an element y is appended to L_1, it computes $\gcd(y, y')$ for all $y' \in L_1$. With probability $\Pr[\text{abort}]$, $\mathscr{A}$ performs a sequence of ring operations such that the list L_1 contains two elements y, y' such that

$$y \not\equiv y' \bmod N \quad \text{and} \quad y \equiv y' \bmod P.$$

In this case, $\gcd(y, y')$ reveals a non-trivial factor P of N.

Clearly, $\mathscr{B}$ succeeds with probability $\Pr[\text{abort}]$, and performs at most t ring operations in $\mathbb{Z}_N$ plus less than t^2 gcd-computations on $\log_2 N$-bit integers. $\square$

3.6 Implications

We can derive the following corollaries from the main theorem from in Section 3.5. The first one confirms a conjecture of Wolf [Wol99, Conjecture 10.1], whereas the second one may be seen as support towards the existence of ring-homomorphic one-way permutations.

Corollary 3.1

If there exists an efficient generic reduction from the discrete logarithm problem to the Diffie-Hellman problem in groups of order N, then there exists an efficient algorithm finding a factor of N.

Note that the above corollary holds even if the reduction algorithm may query an additional "inverting" Diffie-Hellman oracle returning $g^{ab^{-1}}$ on input g^a, g^b. Such an oracle does not to seem to be implied by a Diffie-Hellman oracle, if the order $\phi(N)$ of the group $\mathbb{Z}_N^*$ is unknown. Recall here that computing $\phi(N)$ on input N is as hard as factoring N [Mil76, RSA78, May04].

However, our results say nothing if factoring N is easy, for instance if N is prime. Thus, there may still exist a generic reduction from the discrete logarithm to the Diffie-Hellman problem which works in any group whose order can be factored efficiently.

Another implication is that in general solely the *structure* of a ring is not sufficient to invert a ring-homomorphic one-way permutation efficiently.

Corollary 3.2

If there exists an algorithm inverting any ring-homomorphic one-way permutation with ring structure isomorphic to $\mathbb{Z}_N$ efficiently, then there exists an efficient algorithm finding a factor of N.

3.7 Extensions

The proof of Theorem 3.1 can be generalized from the ring $\mathbb{Z}_N$ to finite commutative unitary rings of characteristic N, by combining the technique presented here with results of [AJR08].

Let us sketch this. Let $N = \prod_{i=1}^k P_i^{e_i}$ be the prime factor decomposition of N, and let R_N be a ring of characteristic N. The extension exploits that any such ring R_N is decomposable into a direct product of rings $R_N \cong R_P \times R_Q$ such that R_P has characteristic P, where $P = p^k$ is a prime power such that $p^k \mid N$, but $p^{k+1} \nmid N$.

Let $\xi : R_N \to R_P \times R_Q$ be the isomorphism mapping R_N to $R_P \times R_Q$. We adopt the proof framework of Theorem 3.1 from $\mathbb{Z}_N$ to R_N as follows.

1. First, the abort condition in Game 1 is modified such that abort is raised if the algorithm computes two ring elements y, y' such that the difference $\delta = y - y'$ corresponds to a ring element $\delta \in R_N$ with

$$\xi(\delta) = (0, \delta_Q) \in R_P \times R_Q.$$

   Thus, we can construct an algorithm finding such ring elements with probability $\Pr[\text{abort}]$.

2. Then in Game 2 we introduce an oracle which uses the subring R_P instead of R_N for the internal representation of ring elements, and which performs all computations in R_P. Thus the success probability of any BBRE algorithm in Game 2 is at most $|R_P|/|R_N|$.

3. Finally, we use the algorithms described in [AJR08] to construct a factoring algorithm which can be used to factorize N efficiently, given a ring element δ with $\xi(\delta) = (0, \delta_Q)$. We refer to [AJR08] for details.

The cryptographic implications of this extension regarding the "DL-vs.-DH"-question are limited, since in cryptography usually cyclic groups are used which are covered by the $\mathbb{Z}_N$-case. However, the above extension indicates that ring-homomorphic one-way permutations not only with a structure isomorphic to $\mathbb{Z}_N$ may exist, but also with a structure isomorphic to arbitrary finite commutative unitary rings of characteristic N.

Corollary 3.3
Let R be a finite commutative unitary ring with characteristic N. If there exists an algorithm inverting any ring-homomorphic one-way permutation with ring structure isomorphic to R efficiently, then there exists an efficient factoring algorithm.

4 Analysis of Cryptographic Assumptions in the Generic Ring Model

One goal of the generic group model is to provide a reasonable abstraction for certain groups, such as elliptic curve groups, for which not many properties beyond the abstractly defined properties of a group are known.

However, important computational problems, such as the RSA problem and the quadratic residuosity problem, are defined over the multiplicative group $(\mathbb{Z}_N^*, \cdot)$, represented by integers modulo N. This representation exhibits many properties beyond the abstract group definition, such as for instance the fact that the group $(\mathbb{Z}_N^*, \cdot)$ is embedded into the ring $(\mathbb{Z}_N, +, \cdot)$. The generic group model seems too restrictive to provide a tool for a meaningful analysis of such problems.

As an approach to reflect the additional algebraic structure of a ring, the notion of generic groups was extended to generic *rings*. A long line of research [BL96, BV98, Bro05, LR06, MR07, AJR08, AM09, AMS11] analyzes cryptographically relevant computational problems and their relationships in the *generic ring model*. Recall from Chapter 2.3 that this model is a simple extension of the generic group model, which allows to compute an additional algebraic operation, such that the resulting structure forms a ring. Clearly, when considering hardness assumptions defined over rings, then this idealization is much more appropiate than the generic group model.

The RSA problem was studied extensively in the generic ring model, see [BV98, Bro05, LR06, AM09, AMS11]. For instance, Aggarwal and Maurer [AM09] show that solving the RSA problem with generic ring algorithms is equivalent to factoring integers. A common conclusion drawn in previous works is that a proof in the generic model supports the conjecture that breaking RSA is also equivalent to factoring integers in a standard model of computation. Is this conclusion reasonable?

4.1 Main Results

We prove a main theorem which states that solving certain subset membership problems in $\mathbb{Z}_N$ with generic ring algorithms is equivalent to factoring N. Then we show that this main theorem has both positive and negative implications.

Negative implications. The main theorem allows us to provide an example of a computational problem of high cryptographic relevance which is equivalent to factoring in the generic model, but easy to solve if elements of $\mathbb{Z}_N$ are given in their standard representation as integers. Concretely, we prove that computing the *Jacobi symbol* [Sho08, Chapter 12.2] of an integer modulo N in the generic ring model is equivalent to factoring N. Since there are simple and efficient non-generic algorithms which compute the Jacobi symbol [Sho08, Chapter 12.2], this provides an example of a natural computational problem which is hard in the generic ring model, but easy to solve if elements of $\mathbb{Z}_N$ are given in their standard representation as integers. Thus, a proof in the generic ring model is unfortunately not a strong indicator for the hardness of a computational problem in the standard model.

Interpretation. For many common idealized models in cryptography it has been shown that a cryptographic reduction in the ideal model need not guarantee security in the "real world". Famous examples are the random oracle model [CGH04], the ideal cipher model [Bla06], and the generic group model [Fis00, Den02]. All these results have in common that they provide somewhat "artificial" computational problems that deviate from standard cryptographic practice.

Note, however, that both the definition and the algebraic properties of the Jacobi symbol are *remarkably* similar to the *quadratic residuosity problem* [GM84], which builds the foundation of numerous cryptosystems and is widely conjectured to be hard. Thus, in contrast to previous works, the equivalence of computing the Jacobi symbol generically and factoring is an example of a *natural* computational problem that is provably hard in the generic model, but easy to solve if elements of $\mathbb{Z}_N$ are given in their standard representation as integers modulo N. This is an important aspect for interpreting results in the generic ring model, like [Bro05, LR06, AM09, AMS11]. Thus, a proof in the generic model is unfortunately not a strong indicator that the considered problem is indeed useful for cryptographic applications.

Positive implications. Despite this negative result, generic hardness results still provide a lower complexity bound for a large class of algorithms, namely all algorithms solving a problem independent of a given representation of ring elements. Motivated by this fact, we show as another application of our main theorem that solving the well-known *quadratic residuosity problem* [GM84] generically is equivalent to factoring. Thus, from a cryptanalytic point of view, we cannot hope to find an algorithm solving this problem efficiently without exploiting any property of the representation of ring elements, unless factoring integers is easy.

4.2 Related Work

Brown [Bro05] reduced the problem of factoring integers to solving the *low-exponent* RSA problem with *straight line programs*, which are a subclass of generic ring algorithms. Damgård and Koprowski showed the generic intractability of root extraction in groups of hidden order [DK02]. Leander and Rupp [LR06] augmented this result to generic ring algorithms, where the considered algorithms may only perform the operations addition, subtraction and multiplication modulo n, but not multiplicative inversions. This result was extended by Aggarwal and Maurer [AM09] from low-exponent RSA to full RSA and to generic ring algorithms that may also compute multiplicative inverses, and by Aggarwal, Maurer and Shparlinski [AMS11] to the related strong RSA problem. Furthermore, Boneh and Venkatesan [BV98] have shown that there is no straight line program reducing integer factorization to the low-exponent RSA problem, unless factoring integers is easy.

The notion of generic ring algorithms has also been applied to study the relationship between the discrete logarithm and the Diffie-Hellman problem, and the existence of ring-homomorphic one-way permutations [BL96, MR07, AJR08].

Comparison to previous works. In contrast to previous work [Bro05, LR06, AJR08, AM09, AMS11], where integer factorization is reduced to solving *search* problems (in the sense that the algorithm has to search for a certain ring element or integer), we show that in order to factor N it suffices to be able to solve *decisional* problems in $\mathbb{Z}_N$. Our results do not only cover the case where N is the product of two primes, but hold in the general case where N is the product of *at least* two primes.

4.3 Definitions

For $\ell \in \mathbb{N}$ we write $[\ell]$ to denote the set $[\ell] = \{1, \ldots, \ell\}$. We denote with $a \xleftarrow{\$} A$ the action of sampling a uniformly random element a from set A. Throughout the chapter we let N be the product of at least two different primes, and denote with $N = \prod_{i=1}^{\ell} p_i^{e_i}$ the prime factor decomposition of N such that $\gcd(p_i, p_j) = 1$ for $i \neq j$. Occasionally we write $a \equiv_N b$ shorthand for $a \equiv b \mod N$.

Let $P = (S_1, \ldots, S_m)$ be a finite sequence. Then $|P|$ denotes its length, i.e., $|P| = m$. For $k \leq m$, we write $P_k \sqsubseteq P$ to denote that P_k is the subsequence of P that consists of the *first* k elements of P, i.e., $P_k = (S_1, \ldots, S_k)$.

4.3.1 Straight Line Programs

A straight line program P over a ring R is an algorithm performing a fixed sequence of ring operations to its input $x \in R$, without branching or looping, that outputs an element $P(x) \in R$.

In the sequel we are interested in straight line programs over the particular ring $R = \mathbb{Z}_N$, where elements are represented by integers. Note that we can not only compute the ring operations addition, subtraction, and multiplication in the ring $\mathbb{Z}_N$, but we also know how to compute *division*, that is, multiplication by multiplicative inverses (if existent), efficiently. In order to make the class of considered algorithms as broad and natural as possible, we therefore include an explicit division operation, though it is generally not explicitly defined for a ring.

The following definition is a simple adaption of [Bro05, Definition 1] to straight line programs that may also compute multiplicative inverses. For our purposes it is sufficient to consider straight-line programs that take as input a *single* ring element $x \in R$, a generalization to algorithms with more input values is straightforward.

Definition 4.1
A *straight line program* P of length m over R is a sequence of tuples

$$P = ((i_1, j_1, \circ_1), \cdots, (i_m, j_m, \circ_m))$$

where $i_k, j_k \in \{-1, \ldots, m\}$ and $\circ_k \in \{+, -, \cdot, /\}$ for $k \in \{1, \ldots, m\}$. The output $P(x)$ of straight line program P on input $x \in R$ is computed as follows.

1. Initialize $L_{-1} := 1 \in R$ and $L_0 := x$.
2. For k from 1 to m do:
   - if $\circ_k = /$ and L_{j_k} is not invertible, then return $\perp$,
   - else set $L_k := L_{i_k} \circ L_{j_k}$.
3. Return $P(x) = L_m$.

We say that each triple $(i, j, \circ) \in P$ is a *SLP-step*.

For notational convenience, for a given straight line program P we will denote with P_k the straight line program given by the sequence of the first k elements of P, with the additional convention that $P_{-1}(x) = 1$ and $P_0(x) = x$ for all $x \in R$.

4.3.2 Uniform Closure

By the Chinese Remainder Theorem, for $N = \prod_{i=1}^{\ell} p_i^{e_i}$ the ring $\mathbb{Z}_N$ is isomorphic to the direct product of rings

$$\mathbb{Z}_{p_1^{e_1}} \times \cdots \times \mathbb{Z}_{p_\ell^{e_\ell}}.$$

Let ϕ be the isomorphism $\phi : \mathbb{Z}_{p_1^{e_1}} \times \cdots \times \mathbb{Z}_{p_\ell^{e_\ell}} \to \mathbb{Z}_N$, and for $\mathscr{C} \subseteq \mathbb{Z}_N$ let

$$\mathscr{C}_i := \{x \bmod p_i^{e_i} : x \in \mathscr{C}\}$$

for all $i \in [\ell]$.

Definition 4.2
We say that $\mathscr{U}[\mathscr{C}] \subseteq \mathbb{Z}_N$ is the *uniform closure* of $\mathscr{C} \subseteq \mathbb{Z}_N$, if

$$\mathscr{U}[\mathscr{C}] = \{y \in \mathbb{Z}_N : y = \phi(x_1 \ldots, x_\ell), x_i \in \mathscr{C}_i \text{ for } i \in [\ell]\}.$$

Example 4.1
Let p_1, p_2 be different primes, $N := p_1 p_2$, and ϕ be the isomorphism $\mathbb{Z}_{p_1} \times \mathbb{Z}_{p_2} \to \mathbb{Z}_N$. For $x \in \mathbb{Z}_N$ let $x_1 := x \bmod p_1$ and $x_2 := x \bmod p_2$. Consider the subset $\mathscr{C} \subseteq \mathbb{Z}_N$ such that

$$\mathscr{C} = \{a, b\} = \{\phi(a_1, a_2), \phi(b_1, b_2)\}.$$

The uniform closure $\mathscr{U}[\mathscr{C}]$ of $\mathscr{C}$ is the set

$$\mathscr{U}[\mathscr{C}] = \{\phi(a_1, a_2), \phi(b_1, b_2), \phi(a_1, b_2), \phi(a_1, b_2)\}.$$

In particular note that $\mathscr{C} \subseteq \mathscr{U}[\mathscr{C}]$, but not necessarily $\mathscr{U}[\mathscr{C}] \subseteq \mathscr{C}$.

Lemma 4.1
Sampling $y \xleftarrow{\$} \mathscr{U}[\mathscr{C}]$ is equivalent to sampling $z_i \xleftarrow{\$} \mathscr{C}_i$ for $i \in [\ell]$ independently and setting

$$y = \psi(z_1, \ldots, z_\ell).$$

The above follows directly from the definition of $\mathscr{U}[\mathscr{C}]$ and the Chinese Remainder Theorem.

4.3.3 Homogeneous Sets

Definition 4.3
We say that a set $\mathscr{C} \subseteq \mathbb{Z}_N$ is *homogeneous*, if for each $i \in [\ell]$ and for each $c \in \mathbb{Z}_{p_i^{e_i}}$ we have that

$$\Pr[x \equiv c \bmod p_i^{e_i} : x \xleftarrow{\$} \mathscr{C}] = \Pr[y \equiv c \bmod p_i^{e_i} : y \xleftarrow{\$} \mathscr{U}[\mathscr{C}]].$$

Putting it differently, $\mathscr{C}$ is homogeneous, if for each $i \in [\ell]$ and for $x \xleftarrow{\$} \mathscr{C}$ and $y \xleftarrow{\$} \mathscr{U}[\mathscr{C}]$ we have that $x \bmod p_i^{e_i}$ is identically distributed to $y \bmod p_i^{e_i}$.

Example 4.2
Again let p_1, p_2 be different primes, $N := p_1 p_2$, ϕ be the isomorphism $\mathbb{Z}_{p_1} \times \mathbb{Z}_{p_2} \to \mathbb{Z}_N$, and for $c \in \{a, b\} \subset \mathbb{Z}_N$ let $c_1 := c \bmod p_1$ and $c_2 := c \bmod p_2$.

- Let $\mathscr{C} = \{\phi(a_1,a_2), \phi(a_1,b_2), \phi(b_1,b_2), \phi(b_1,a_2)\}$, then $\mathscr{C} = \mathscr{U}[\mathscr{C}]$. Clearly $\mathscr{C} = \mathscr{U}[\mathscr{C}]$ implies that $\mathscr{C}$ is homogeneous.
- Let $\mathscr{C}' = \{\phi(a_1,a_2), \phi(a_1,b_2), \phi(b_1,b_2)\}$, then we have $\mathscr{U}[\mathscr{C}'] = \mathscr{C}$. Note that it holds that

$$\Pr[x \equiv a_1 \bmod p_1 : x \xleftarrow{\$} \mathscr{C}'] = 2/3,$$

but we have

$$\Pr[x \equiv a_1 \bmod p_1 : x \xleftarrow{\$} \mathscr{U}[\mathscr{C}']] = 1/2.$$

Thus, $\mathscr{C}'$ is not homogeneous.

- Let $\mathscr{C}'' = \{\phi(a_1,a_2), \phi(b_1,b_2)\}$, then again we have $\mathscr{U}[\mathscr{C}''] = \mathscr{C}$. $\mathscr{C}''$ is homogeneous, since we have

$$\Pr[x \equiv c_i \bmod p_i : x \xleftarrow{\$} \mathscr{C}''] = 1/2 = \Pr[x \equiv c_i \bmod p_i : x \xleftarrow{\$} \mathscr{U}[\mathscr{C}'']]$$

for all $i \in \{1,2\}$ and $c_i \in \{a_i, b_i\}$.

4.4 Straight Line Programs over the Ring $\mathbb{Z}_N$

In the following we will state a few lemmas on straight line programs over $\mathbb{Z}_N$ that will be useful for the proof of our main result.

Lemma 4.2
Suppose there exists a straight line program P such that for $x, x' \in \mathbb{Z}_N$ holds that

$$P(x') \neq \bot \quad \text{and} \quad P(x) = \bot.$$

Then there exists $P_j \sqsubseteq P$ such that

$$P_j(x') \in \mathbb{Z}_N^* \quad \text{and} \quad P_j(x) \in \mathbb{Z}_N \setminus \mathbb{Z}_N^*.$$

PROOF. $P(x) = \bot$ means that there exists an SLP-step $(i,j,\circ) \in P$ such that $\circ = /$ and $L_j = P_j(x) \in \mathbb{Z}_N \setminus \mathbb{Z}_N^*$. However, $P(x')$ does not evaluate to $\bot$, thus it must hold that $P_j(x') \in \mathbb{Z}_N^*$. $\qquad\square$

The following lemma provides a lower bound on the probability of factoring N by evaluating a straight line program P with a random value $y \xleftarrow{\$} \mathscr{U}[\mathscr{C}]$ and computing $\gcd(N, P(y))$, relative to the probability that $P(x') \in \mathbb{Z}_N \setminus \mathbb{Z}_N^*$ and $P(x) \in \mathbb{Z}_N^*$ for randomly chosen $x, x' \xleftarrow{\$} \mathscr{C}$.

Lemma 4.3
Let $N = \prod_{i=1}^{\ell} p_i^{e_i}$ with $\ell \geq 2$, and let $\mathscr{C} \subseteq \mathbb{Z}_N$ be homogeneous. For any straight line program P, and for uniformly random $x, x' \xleftarrow{\$} \mathscr{C}$ and $y \xleftarrow{\$} \mathscr{U}[\mathscr{C}]$, holds that

$$\Pr\left[P(x') \in \mathbb{Z}_N n \setminus \mathbb{Z}_N^* \text{ and } P(x) \in \mathbb{Z}_N^*\right] \leq \Pr\left[\gcd(N, P(y)) \notin \{1, N\}\right].$$

Similar to the above, the following lemma provides a lower bound on the probability of factoring N by computing $\gcd(N, P(y) - Q(y))$ with $y \xleftarrow{\$} \mathscr{U}[\mathscr{C}]$ for two given straight line programs P and Q, relative to the probability that $P(x) \equiv_N Q(x)$ and $P(x') \not\equiv_N Q(x')$ for random $x, x' \xleftarrow{\$} \mathscr{C}$.

Lemma 4.4
Let $N = \prod_{i=1}^{\ell} p_i^{e_i}$ with $\ell \geq 2$, and let $\mathscr{C} \subseteq \mathbb{Z}_N$ be homogeneous. For any pair (P, Q) of straight line programs, $x, x' \xleftarrow{\$} \mathscr{C}$, and $y \xleftarrow{\$} \mathscr{U}[\mathscr{C}]$ holds that

$$\Pr\left[P(x) \equiv_N Q(x) \text{ and } P(x') \not\equiv_N Q(x')\right] \leq \Pr\left[\gcd(N, P(y) - Q(y)) \notin \{1, N\}\right].$$

Before proving Lemmas 4.3 and 4.4, we will give some intuition in the following section.

4.4.1 Some Intuition for Lemma 4.3 and 4.4

Simplifying a little, Lemma 4.3 and 4.4 state *essentially* that: if we are given a straight line program mapping "many" inputs to zero and "many" inputs to a non-zero value, then we can find a factor of N by sampling $y \xleftarrow{\$} \mathscr{U}[\mathscr{C}]$ and computing $\gcd(N, P(y))$.[1] At a first glance this seems counterintuitive.

The simple case: $\mathscr{C} = \mathbb{Z}_N$. As an example let us consider the case $\mathscr{C} = \mathbb{Z}_N$ with $N = p_1 p_2$, where p_1 and p_2 are not necessarily prime, but $p_1, p_2 > 1$ and $\gcd(p_1, p_2) = 1$. Note that we have $\mathscr{U}[\mathscr{C}] = \mathbb{Z}_N$. Assume a straight line program P mapping about one half of the elements of $\mathbb{Z}_N$ to 0, and the other half to 1. Then P maps "many" inputs to zero and "many" inputs to a non-zero value, but clearly computing $\gcd(N, P(y))$ for any $y \in \mathbb{Z}_N$ yields only trivial factors of N. This seems to be a counterexample to Lemma 4.3 and 4.4. However, in fact it is not, since there exists no straight line program P satisfying the assumed property, if N is the product of at least two different primes.

---

[1] In case of Lemma 4.3 note that $P(x) \in \mathbb{Z}_N^*$ and $P(x') \in \mathbb{Z}_N \setminus \mathbb{Z}_N^*$ means that $P(x')$ is zero modulo *at least one* prime factor of N, while $P(x) \not\equiv 0$ modulo *all* prime factors of N. In case of Lemma 4.4 observe that if we have $P(x) - Q(x) \equiv 0 \bmod N$ and $P(x') - Q(x') \not\equiv 0 \bmod N$, then x is mapped to zero and x' is not mapped to zero by the straight line program $S(x) := P(x) - Q(x)$.

The reason for this is a consequence of the Chinese Remainder Theorem, which states that the ring $\mathbb{Z}_N$ is isomorphic to $\mathbb{Z}_{p_1} \times \mathbb{Z}_{p_2}$. Let $\phi : \mathbb{Z}_{p_1} \times \mathbb{Z}_{p_2} \to \mathbb{Z}_N$ denote this isomorphism. Assume $x, x' \in \mathbb{Z}_N$ and a straight line program P such that $P(x) \equiv 0 \bmod N$ and $P(x') \equiv 1 \bmod N$. Since ϕ is a ring-isomorphism and P performs only ring operations, it holds that

$$P(x) = \phi(P(x) \bmod p_1, P(x) \bmod p_2) = \phi(0,0)$$

and

$$P(x') = \phi(P(x') \bmod p_1, P(x') \bmod p_2) = \phi(1,1).$$

The crucial observation is now that for each pair $(x, x') \in \mathbb{Z}_N^2$, there exist $c, d \in \mathbb{Z}_N$ such that $c = \phi(x' \bmod p_1, x \bmod p_2)$ and $d = \phi(x \bmod p_1, x' \bmod p_2)$. Evaluating P with c or d yields

$$P(c) = \phi(P(x') \bmod p_1, P(x) \bmod p_2) = \phi(1,0)$$

or

$$P(d) = \phi(P(x) \bmod p_1, P(x') \bmod p_2) = \phi(0,1).$$

We therefore have $\gcd(N, P(c)) = p_2$ and $\gcd(N, P(d)) = p_1$.

In this example we have $\mathscr{C} = \mathscr{U}[\mathscr{C}] = \mathbb{Z}_N$, and we assume that P has the property that $P(x) = \phi(0,0)$ and $P(x') = \phi(1,1)$ with "high" probability for *uniformly random* $x, x' \xleftarrow{\$} \mathbb{Z}_N$. The crucial observation is now that the Chinese Remainder Theorem implies that if we sample $y \xleftarrow{\$} \mathbb{Z}_N$ uniformly random, then we also have with "high" probability that $P(y) = \phi(0,1)$ or $P(y) = \phi(1,0)$. A factor of N can therefore be found by sampling y and computing $\gcd(N, P(y))$.

The general case: $\mathscr{C} \subset \mathbb{Z}_N$. The proofs of Lemma 4.3 and 4.4 generalize the above idea to the case where $\mathscr{C}$ is a *subset* of $\mathbb{Z}_N$. This generalization made it necessary to define the *uniform closure* $\mathscr{U}[\mathscr{C}]$ and *homogeneous sets*.

For instance, consider a subset $\mathscr{C} = \{x, x'\}$ with $x = \phi(x_p, x_q)$ and $x' = \phi(x'_p, x'_q)$. Suppose we are given a straight line program P such that $P(x) = 0$ and $P(x') = 1$. We can factor N using this straight line program by computing $\gcd(N, P(y))$, if we can find a suitable y such that $y \in \{\phi(x_p, x'_q), \phi(x'_p, x_q)\}$.

The uniform closure $\mathscr{U}[\mathscr{C}]$ is defined such that we know that it contains such a suitable $y \in \mathscr{U}[\mathscr{C}]$. Moreover, as we show in the proofs of Lemma 4.3 and 4.4, if $\mathscr{C}$ is homogeneous, then we can find a suitable y with sufficiently high probability simply by sampling $y \xleftarrow{\$} \mathscr{U}[\mathscr{C}]$ uniformly random.

Finally, in order to obtain an *efficient* factoring algorithm, we will need to require that there exist efficient sampling algorithms for $\mathscr{C}$ and $\mathscr{U}[\mathscr{C}]$. We will have to show this separately for each considered subset membership problem.

4.4.2 Proof of Lemma 4.3

First, observe that $P(x') \in \mathbb{Z}_N \setminus \mathbb{Z}_N^*$ implies, that there exists at least one $i \in [\ell]$ such that $P(x') \equiv 0 \bmod p_i$, while $P(x) \in \mathbb{Z}_N^*$ implies that $P(x) \in \mathbb{Z}_{p_j}^*$ for all $j \in [\ell]$. Thus, we have

$$\Pr[P(x') \in \mathbb{Z}_N \setminus \mathbb{Z}_N^* \text{ and } P(x) \in \mathbb{Z}_N^*]$$
$$= \Pr[\exists i \in [\ell] \text{ s.t. } P(x') \equiv 0 \bmod p_i \text{ and } P(x) \in \mathbb{Z}_{p_j}^* \text{ for all } j \in [\ell]]$$
$$\leq \Pr[\exists i, j \in [\ell] \text{ s.t. } j \neq i \text{ and } P(x') \equiv 0 \bmod p_i \text{ and } P(x) \in \mathbb{Z}_{p_j}^*].$$

Note furthermore that we have $P(x) \equiv P(x \bmod p_i^{e_i}) \bmod p_i$, since P performs only ring operations. Thus we have

$$\Pr[\exists i, j \in [\ell] \text{ s.t. } j \neq i \text{ and } P(x') \equiv 0 \bmod p_i \text{ and } P(x) \in \mathbb{Z}_{p_j}^*]$$
$$= \Pr[\exists i, j \in [\ell] \text{ s.t. } j \neq i \text{ and } P(x_i') \equiv 0 \bmod p_i \text{ and } P(x_j) \in \mathbb{Z}_{p_j}^*],$$

where $x_i' := x' \bmod p_i^{e_i}$ and $x_j := x \bmod p_j^{e_j}$.

Since $\mathscr{C}$ is homogeneous, we have that sampling $x, x' \xleftarrow{\$} \mathscr{C}$ and computing $x_i' = x' \bmod p_i^{e_i}$ and $x_j = x \bmod p_j^{e_j}$ is equivalent to sampling $z, z' \xleftarrow{\$} \mathscr{U}[\mathscr{C}]$ and setting $z_i' := z' \bmod p_i^{e_i}$ and $z_j := z \bmod p_j^{e_j}$. Thus we have

$$\Pr[\exists i, j \in [\ell] \text{ s.t. } j \neq i \text{ and } P(x_i') \equiv 0 \bmod p_i \text{ and } P(x_j) \in \mathbb{Z}_{p_j}^*]$$
$$= \Pr[\exists i, j \in [\ell] \text{ s.t. } j \neq i \text{ and } P(z_i') \equiv 0 \bmod p_i \text{ and } P(z_j) \in \mathbb{Z}_{p_j}^*]$$

for $z, z' \xleftarrow{\$} \mathscr{U}[\mathscr{C}]$.

Now Lemma 4.1 states that for $z, z', y \xleftarrow{\$} \mathscr{U}[\mathscr{C}]$ holds that

$$\Pr[\exists i, j \in [\ell] \text{ s.t. } j \neq i \text{ and } P(z_i') \equiv 0 \bmod p_i \text{ and } P(z_j) \in \mathbb{Z}_{p_j}^*]$$
$$= \Pr[\exists i, j \in [\ell] \text{ s.t. } j \neq i \text{ and } P(y_i) \equiv 0 \bmod p_i \text{ and } P(y_j) \in \mathbb{Z}_{p_j}^*],$$

where $y_i = y \bmod p_i^{e_i}$ and $y_j = y \bmod p_j^{e_j}$.

Using again that P performs only ring operations, we obtain that

$$\Pr[\exists i, j \in [\ell] \text{ s.t. } j \neq i \text{ and } P(y_i) \equiv 0 \bmod p_i \text{ and } P(y_j) \in \mathbb{Z}_{p_j}^*]$$
$$= \Pr[\exists i, j \in [\ell] \text{ s.t. } j \neq i \text{ and } P(y) \equiv 0 \bmod p_i \text{ and } P(y) \in \mathbb{Z}_{p_j}^*].$$

Finally, we find a factor of N by computing $\gcd(N, P(y))$ if there exists $i, j \in [\ell]$ such that $P(y) \equiv 0 \bmod p_i$ and $P(y) \in \mathbb{Z}_{p_j}^*$. Thus we have

$$\Pr[\exists i, j \in [\ell] \text{ s.t. } j \neq i \text{ and } P(y) \equiv 0 \bmod p_i \text{ and } P(y) \in \mathbb{Z}_{p_j}^*]$$
$$\leq \Pr[\gcd(N, P(y)) \notin \{1, n\}].$$

4.4.3 Proof of Lemma 4.4

Let $x, x' \xleftarrow{\$} \mathscr{C}$ and $y \xleftarrow{\$} \mathscr{U}[\mathscr{C}]$, and let us write $a_i := a \bmod p_i^{e_i}$ for all $a \in \{x', x, y\}$. Let $\Delta(x) := P(x) - Q(x)$. Then, with the same arguments as in the proof of Lemma 4.3, we have

$$\Pr\left[\Delta(x') \not\equiv_N 0 \text{ and } \Delta(x) \equiv_N 0\right]$$
$$= \Pr\left[\exists i \text{ s.t. } \Delta(x') \not\equiv 0 \bmod p_i^{e_i} \text{ and } \Delta(x) \equiv 0 \bmod p_j^{e_j} \text{ for all } j \in [\ell]\right]$$
$$\leq \Pr\left[\exists i, j \text{ s.t. } j \neq i \text{ and } \Delta(x') \not\equiv 0 \bmod p_i^{e_i} \text{ and } \Delta(x) \equiv 0 \bmod p_j^{e_j}\right]$$
$$= \Pr\left[\exists i, j \text{ s.t. } j \neq i \text{ and } \Delta(x'_i) \not\equiv 0 \bmod p_i^{e_i} \text{ and } \Delta(x_j) \equiv 0 \bmod p_j^{e_j}\right]$$
$$= \Pr\left[\exists i, j \text{ s.t. } j \neq i \text{ and } \Delta(y_i) \not\equiv 0 \bmod p_i^{e_i} \text{ and } \Delta(y_j) \equiv 0 \bmod p_j^{e_j}\right]$$
$$= \Pr\left[\exists i, j \text{ s.t. } j \neq i \text{ and } \Delta(y) \not\equiv 0 \bmod p_i^{e_i} \text{ and } \Delta(y) \equiv 0 \bmod p_j^{e_j}\right]$$
$$\leq \Pr[\gcd(N, \Delta(y)) \notin \{1, N\}].$$

4.5 Subset Membership Problems in the Generic Ring Model

Definition 4.4
Let $\mathscr{C} \subseteq \mathbb{Z}_N$ and $\mathscr{V} \subseteq \mathscr{C}$ with $|\mathscr{V}| > 1$. The *subset membership problem* defined by $(\mathscr{C}, \mathscr{V})$ is: given $x \xleftarrow{\$} \mathscr{C}$, decide whether $x \in \mathscr{V}$.

In this chapter we will consider only subset membership problems such that $|\mathscr{C}| = 2 \cdot |\mathscr{V}|$, this is sufficient for all our applications.

We formalize the notion of subset membership problems in the generic ring model as a game between an algorithm $\mathscr{A}$ and a generic ring oracle $\mathscr{O}$. Oracle $\mathscr{O}$ is defined exactly like the generic ring oracle described in Section 2.3, and receives $1 \in \mathbb{Z}_N$ and a uniformly random element $x \xleftarrow{\$} \mathscr{C}$ as input. That is, the adversary interacts with an oracle

$$\mathcal{O}(\mathbb{Z}_N, \Pi, \Sigma; (1, x)),$$

where $x \xleftarrow{\$} \mathscr{C}$ is chosen uniformly random from $\mathscr{C}$, $\Pi = \{+, -, \cdot, \div\}$, and $\Sigma = \{=\}$.

Let $\mathsf{Succ}(\mathscr{A})$ denote the event that an algorithm $\mathscr{A}$ interacting with $\mathcal{O}$ solves the given instance of the subset membership problem successfully. That is, $\mathsf{Succ}(\mathscr{A})$ occurs if

$$\mathscr{A}^{\mathcal{O}(\mathbb{Z}_N, \Pi, \Sigma; (1, x))}(N) = 1 \text{ and } x \in \mathscr{V}, \quad \text{or} \quad \mathscr{A}^{\mathcal{O}(\mathbb{Z}_N, \Pi, \Sigma; (1, x))}(N) = 0 \text{ and } x \notin \mathscr{V}.$$

Note that any algorithm for a given subset membership problem $(\mathscr{C}, \mathscr{V})$ has at least the trivial success probability $1/2$ of solving a given problem instance correctly by guessing. Note also that $\Pr[x \in \mathscr{V} : x \xleftarrow{\$} \mathscr{C}] = 1/2$, since x is chosen uniformly random and we have $|\mathscr{C}| = 2 \cdot |\mathscr{V}|$.

Definition 4.5
We say that a generic ring algorithm $\mathscr{A}$ (ε, t)-solves the subset membership problem $(\mathscr{C}, \mathscr{V})$, if $\mathscr{A}$ issues at most t oracle queries, and

$$\Pr[\mathsf{Succ}(\mathscr{A})] \geq |1/2 + \varepsilon|.$$

4.5.1 Implementing the Generic Ring Oracle

For our results presented in the sequel, it will be helpful to have an abstract "implementation" of the generic ring oracle $\mathcal{O}$ described above. That is, we will describe some specific details how the oracle $\mathcal{O}$ processes queries internally. Later in the proof we will modify these procedures.

Let us define the following two procedures.

- The Compute-procedure takes as input two indices (i, j) and a symbol $\circ \in \{+, -, \cdot, /\}$ as input. The procedure returns false if $\circ = /$ and $L_j \notin \mathbb{Z}_N^*$. Otherwise it computes $L_i \circ L_j \bmod N$, appends the result to L, and returns true.

- The Equal-procedure takes two indices (i, j) as input. The procedure returns true if $L_i \equiv L_j \bmod N$ and false otherwise.

Whenever the algorithm submits a query $(i, j, \circ)$ with $\circ \in \{+, -, \cdot, \div\}$, the oracle runs Compute$(i, j, \circ)$, and returns $\perp$ if Compute returns false. Whenever the algorithm makes a query $(i, j, =)$, the oracle returns Equal(i, j).

4.6 Main Result

Our main result relates the probability that an algorithm $\mathscr{A}$ solves an instance of a given subset membership problem $(\mathscr{C}, \mathscr{V})$ to the probability of factoring N with an algorithm $\mathscr{B}$ that runs $\mathscr{A}$ as a subroutine by simulating the generic ring oracle for $\mathscr{A}$.

Theorem 4.1

Let $N = \prod_{i=1}^{\ell} p_i^{e_i}$. Let $(\mathscr{C}, \mathscr{V})$ be a subset membership problem over $\mathbb{Z}_N$ such that $\mathscr{C}$ is homogeneous and $2 \cdot |\mathscr{V}| = |\mathscr{C}|$. For any generic ring algorithm $\mathscr{A}$ (ε, t)-solving $(\mathscr{C}, \mathscr{V})$, there exists an algorithm $\mathscr{B}$ that outputs a non-trivial factor of N with success probability at least

$$\frac{\varepsilon}{2(t^2 + 4t + 3)}$$

by running $\mathscr{A}$ once, performing at most $2t$ additional operations in $\mathbb{Z}_N$ and at most $(t+2)^2$ gcd-computations on $\lceil \log_2 N \rceil$-bit numbers, and sampling each one random element from $\mathscr{C}$ and $\mathscr{U}[\mathscr{C}]$.

Note that the factoring algorithm $\mathscr{B}$ from the above theorem is efficient *only if* we can efficiently sample uniformly random elements from $\mathscr{C}$ and $\mathscr{U}[\mathscr{C}]$. In general such an algorithm need not exist for any subset $\mathscr{C} \subseteq \mathbb{Z}_N$. There are simple examples for sets $\mathscr{C}$ where sampling uniformly random from $\mathscr{U}[\mathscr{C}]$ is already equivalent to factoring.[2] Thus, in order to apply the above theorem to show that factoring reduces efficiently to solving a given subset membership problem $(\mathscr{C}, \mathscr{V})$ in the generic ring model, we will also have to show that there exists efficient sampling algorithms for $\mathscr{C}$ and $\mathscr{U}[\mathscr{C}]$.

Proof outline. We proceed in a short sequence of games (cf. [Sho04]). We start in Game 0 with the original game played between algorithm $\mathscr{A}$ and the generic ring oracle $\mathscr{O}$ described above. Then in Game 1 we replace oracle $\mathscr{O}$ with an oracle $\mathscr{O}_1$. This oracle uses slightly modified procedures $\texttt{Compute}_1$ and $\texttt{Equal}_1$, but is equivalent to $\mathscr{O}$, and thus perfectly indistinguishable from $\mathscr{O}$ for $\mathscr{A}$. In Game 2 $\mathscr{O}_1$ is then replaced with an oracle $\mathscr{O}_2$, which uses procedures $\texttt{Compute}_2$ and $\texttt{Equal}_2$ such that all computations are performed *independent* of the ring element x. Thus, in this game the success probability of $\mathscr{A}$ equals the trivial success probability $1/2$.

---

[2]For instance, if $N = pq$ is the product of two primes and $\mathscr{C} := \{0, 1\} \subset \mathbb{Z}_N$, then the uniform closure of $\mathscr{C}$ is equal to $\mathscr{U}[\mathscr{C}] = \{0, 1, p(p^{-1} \bmod q), q(q^{-1} \bmod p)\}$. Clearly computing $\gcd(N, p(p^{-1} \bmod q)) = p$ or $\gcd(N, q(q^{-1} \bmod p)) = q$ reveals a non-trivial factor of N.

Clearly, any algorithm $\mathscr{A}$ which has a success probability significantly better than the trivial guessing-algorithm must be able to distinguish $\mathcal{O}_2$ from $\mathcal{O}_1$. However, this is possible only if $\mathcal{O}_2$ fails to simulate $\mathcal{O}_1$ perfectly. We thus conclude by showing that there exists a factoring algorithm $\mathscr{B}$ which runs $\mathscr{A}$ as a subroutine and returns a factor of N, such that the success probability of $\mathscr{B}$ corresponds (essentially) to the probability that $\mathcal{O}_2$ fails to simulate $\mathcal{O}_1$.

Remark. The idea of making the computations of the generic ring algorithm independent of the challenge input value was introduced by Gregor Leander and Andy Rupp [LR06, Lemma 2] for the case where $N = pq$ is the product of two primes, $\mathscr{C} = \mathbb{Z}_N$, and generic ring algorithms that do not compute multiplicative inverses. This was generalized in [Jag07, Chapter 5.3.2] and independently in [AM09, Lemma 7] to generic ring algorithms that may also compute inverses, still for the case $\mathscr{C} = \mathbb{Z}_N$ with N the product of two primes. To prove our theorem we have to generalize this to the general case where algorithms may compute inverses, $N = \prod_{i=1}^{\ell} p_i^{e_i}$ is the product of at least two different primes, and where $\mathscr{C} \subseteq \mathbb{Z}_N$ may be a subset of $\mathbb{Z}_N$.

4.6.1 Sequence of Games

Let $\mathsf{Succ}_i(\mathscr{A})$ denote the event that $\mathscr{A}$ solves the given instance of the subset membership problem successfully in Game i.

Game 0. In this game the algorithm $\mathscr{A}$ interacts with the generic ring oracle $\mathcal{O}(\mathbb{Z}_N, \Pi, \Sigma; (1, x))$, where $x \xleftarrow{\$} \mathscr{C}$ is chosen uniformly random from $\mathscr{C}$. We assume that $\mathcal{O}$ internally uses the procedures Compute and Equal, as described above. Clearly, we have

$$\Pr[\mathsf{Succ}_0(\mathscr{A})] \geq 1/2 + \varepsilon$$

for some $\varepsilon \geq 0$.

Game 1. We replace the original oracle $\mathcal{O}$ with an equivalent oracle $\mathcal{O}_1$. Oracle $\mathcal{O}_1$ proceeds exactly like $\mathcal{O}$, except for the following.

Instead of list L, $\mathcal{O}_1$ maintains a sequence P storing the sequence of computations performed by $\mathscr{A}$. P is initialized to the empty sequence. Recall that for $k \geq 1$ we denote with $P_k \sqsubseteq P$ the subsequence of P consisting of the first k entries of P, and with $P_k(x)$ the evaluation of P_k on input x, and that we have $P_{-1}(x) = 1$ and $P_0(x) = x$ by Definition 4.1. Due to our construction of the generic ring oracle (newly computed ring elements are always *appended* to L, see Chapter 2) we have

$$P_{k-2}(x) = L_k$$

for all $k \in \{1, \ldots, t + 2\}$. Thus, it does not matter whether the oracle uses L_k or $P_{k-2}(x)$ in order to perform computations or equality tests. Therefore we replace the Compute and Equal procedures with the following procedures:

- Compute_1: Given a triple $(i, j, \circ) \in \{1, \ldots, t\} \times \{+, -, \cdot, \div\}$, the Compute_1 procedure returns false if $\circ = \div$ and $P_{j-2}(x) \in \mathbb{Z}_N \setminus \mathbb{Z}_N^*$. Otherwise $(i, j, \circ)$ is appended to P, and true is returned.

- Equal_1: The Equal_1-procedure takes a tuple $(i, j) \in \{1, \ldots, t\} \times \{1, \ldots, t\}$ as input. The procedure returns true if $P_{i-2}(x) \equiv P_{j-2}(x) \mod n$ and false otherwise.

Note that the only differences between $\mathcal{O}$ and $\mathcal{O}_1$ are that

- $\mathcal{O}_1$ records the sequence of computations issued by $\mathscr{A}$ in the sequence P, instead of applying these computations directly to elements of the list L, and

- instead of testing whether some list element L_j is invertible (resp. whether two list elements L_i and L_j are equal), $\mathcal{O}_1$ recomputes the required ring elements first, and checks then whether $P_{i-2}(x)$ is invertible (resp. whether $P_{i-2}(x) \equiv P_{j-2}(x) \mod N$), which is equivalent.

Therefore $\mathcal{O}_1$ is just a different "implementation" of oracle $\mathcal{O}$, but both oracles are perfectly indistinguishable for the adversary, and therefore equivalent. Thus we have

$$\Pr[\text{Succ}_1(\mathscr{A})] = \Pr[\text{Succ}_0(\mathscr{A})].$$

Game 2. We replace oracle $\mathcal{O}_1$ with oracle $\mathcal{O}_2$. Our goal is to make an interaction of $\mathscr{A}$ with its oracle *independent* of the challenge value x. To this end, note that $\mathcal{O}_1$ uses x only inside the Compute_1 and Equal_1 procedures. Let us therefore consider an oracle $\mathcal{O}_2$ which is defined exactly like $\mathcal{O}_2$, but samples $x' \xleftarrow{\$} \mathscr{C}$ at random at the beginning of the game, and replaces the procedures Compute_1 and Equal_1 with procedures Compute_2 and Equal_2.

- Compute_2: Given a triple $(i, j, \circ) \in \{1, \ldots, t\} \times \{+, -, \cdot, \div\}$, Compute_2 returns false if $\circ = \div$ and $P_{j-2}(x') \in \mathbb{Z}_N \setminus \mathbb{Z}_N^*$. Otherwise it appends $(i, j, \circ)$ to P and returns true.

- Equal_2: The Equal_2-procedure takes a tuple $(i, j) \in \{1, \ldots, t\} \times \{1, \ldots, t\}$ as input. The procedure returns true if $P_{i-2}(x') \equiv P_{j-2}(x') \mod N$ and false otherwise.

Note that all computations of $\mathscr{A}$ are independent of the challenge value x when interacting with $\mathscr{O}_2$. Hence, any algorithm $\mathscr{A}$ has at most trivial success probability in the simulation game, and therefore we have

$$\Pr[\mathsf{Succ}_2(\mathscr{A})] = 1/2.$$

We say that a *simulation failure*, denoted $\mathscr{F}$, occurs if $\mathscr{O}_2$ does not simulate $\mathscr{O}_1$ perfectly. Clearly we have

$$\Pr[\mathsf{Succ}_1(\mathscr{A})] \leq \Pr[\mathsf{Succ}_2(\mathscr{A})] + \Pr[\mathscr{F}]$$

by Shoup's Difference Lemma [Sho04, Lemma 1].

4.6.2 Bounding the Probability of a Simulation Failure

Observe that an interaction of $\mathscr{A}$ with $\mathscr{O}_2$ is perfectly indistinguishable from an interaction with $\mathscr{O}_1$, unless at least one of the following events occurs.

1. The $\mathsf{Compute}_2$-procedure fails to simulate $\mathsf{Compute}_1$ perfectly. This means that $\mathsf{Compute}_2$ returns $\mathtt{true}$ on a procedure call where $\mathsf{Compute}_1$ would have returned $\mathtt{false}$, or $\mathsf{Compute}_2$ returns $\mathtt{false}$ where $\mathsf{Compute}_1$ would have returned $\mathtt{true}$. Let $\mathscr{F}_{\mathsf{test}}$ denote the event that this happens on at least one call of $\mathsf{Compute}_2$.

2. The Equal_2-procedure fails to simulate Equal_1 perfectly. This means that Equal_2 has returned $\mathtt{true}$ where Equal_1 would have returned $\mathtt{false}$, or Equal_2 has returned $\mathtt{false}$ where Equal_1 would have returned $\mathtt{true}$. Let $\mathscr{F}_{\mathsf{equal}}$ denote the event that this happens at at least one call of Equal_2.

Since $\mathscr{F}$ implies that at least one of the events $\mathscr{F}_{\mathsf{test}}$ and $\mathscr{F}_{\mathsf{equal}}$ has occurred, it holds that

$$
\begin{aligned}
\Pr[\mathscr{F}] &= \Pr[\mathscr{F}_{\mathsf{test}} \cup \mathscr{F}_{\mathsf{equal}}] \\
&= \Pr[\mathscr{F}_{\mathsf{test}}] + \Pr[\mathscr{F}_{\mathsf{equal}}] - \Pr[\mathscr{F}_{\mathsf{test}} \cap \mathscr{F}_{\mathsf{equal}}] \\
&= \Pr[\mathscr{F}_{\mathsf{test}}] + \Pr[\mathscr{F}_{\mathsf{equal}}] - \Pr[\mathscr{F}_{\mathsf{equal}} \mid \mathscr{F}_{\mathsf{test}}] \cdot \Pr[\mathscr{F}_{\mathsf{test}}] \\
&= \Pr[\mathscr{F}_{\mathsf{test}}] + \Pr[\mathscr{F}_{\mathsf{equal}} \mid \mathscr{F}_{\mathsf{test}}] \cdot \Pr[\mathscr{F}_{\mathsf{test}}] + \Pr[\mathscr{F}_{\mathsf{equal}} \mid \neg\mathscr{F}_{\mathsf{test}}] \cdot \Pr[\neg\mathscr{F}_{\mathsf{test}}] \\
&\quad - \Pr[\mathscr{F}_{\mathsf{equal}} \mid \mathscr{F}_{\mathsf{test}}] \cdot \Pr[\mathscr{F}_{\mathsf{test}}] \\
&= \Pr[\mathscr{F}_{\mathsf{test}}] + \Pr[\mathscr{F}_{\mathsf{equal}} \mid \neg\mathscr{F}_{\mathsf{test}}] \cdot \Pr[\neg\mathscr{F}_{\mathsf{test}}] \\
&\leq \Pr[\mathscr{F}_{\mathsf{test}}] + \Pr[\mathscr{F}_{\mathsf{equal}} \mid \neg\mathscr{F}_{\mathsf{test}}]
\end{aligned}
$$

In the following we will bound $\Pr[\mathscr{F}_{\mathsf{test}}]$ and $\Pr[\mathscr{F}_{\mathsf{equal}} \mid \neg\mathscr{F}_{\mathsf{test}}]$ separately.

Bounding the Probability of $\mathscr{F}_{\text{test}}$. By construction of oracle $\mathcal{O}_2$, $\mathscr{F}_{\text{test}}$ occurs only if $\texttt{Compute}_2$ has returned $\texttt{false}$ where $\texttt{Compute}_1$ would have returned $\texttt{true}$, or vice versa. This happens only if there exists $P_j \sqsubseteq P$ such that

$$(P_j(x) \in \mathbb{Z}_N^* \text{ and } P_j(x') \notin \mathbb{Z}_N^*) \text{ or } (P_j(x') \in \mathbb{Z}_N^* \text{ and } P_j(x) \notin \mathbb{Z}_N^*).$$

Note that we may have $P_j(\tilde{x}) = \perp \notin \mathbb{Z}_N^*$ for $\tilde{x} \in \{x, x'\}$.

We can simplify our analysis a little by applying Lemma 4.2. The existence of $P_j \sqsubseteq P$ such that $P_j(\tilde{x}) = \perp$ implies the existence of $P_k \sqsubseteq P_j$ such that $P_k(\tilde{x}) \in \mathbb{Z}_N \setminus \mathbb{Z}_N^*$. Hence, $\mathscr{F}_{\text{test}}$ occurs only if there exists $P_j \sqsubseteq P$ such that

$$(P_j(x) \in \mathbb{Z}_N^* \text{ and } P_j(x') \in \mathbb{Z}_N \setminus \mathbb{Z}_N^*) \text{ or } (P_j(x') \in \mathbb{Z}_N^* \text{ and } P_j(x) \in \mathbb{Z}_N \setminus \mathbb{Z}_N^*).$$

Note that for one *fixed* P_j we have

$$\Pr\left[(P_j(x') \in \mathbb{Z}_N \setminus \mathbb{Z}_N^* \text{ and } P_j(x) \in \mathbb{Z}_N^*) \text{ or } (P_j(x) \in \mathbb{Z}_N \setminus \mathbb{Z}_N^* \text{ and } P_j(x') \in \mathbb{Z}_N^*)\right]$$
$$\leq 2 \cdot \Pr\left[P_j(x) \in \mathbb{Z}_N \setminus \mathbb{Z}_N^* \text{ and } P_j(x') \in \mathbb{Z}_N^*\right].$$

Thus, by taking the maximum probability over all P_j, we get

$$\Pr[\mathscr{F}_{\text{test}}] \leq 2 \cdot \sum_{j=0}^{t} \Pr\left[P_j(x) \in \mathbb{Z}_N \setminus \mathbb{Z}_N^* \text{ and } P_j(x') \in \mathbb{Z}_N^* : x, x' \xleftarrow{\$} \mathscr{C}\right]$$
$$\leq 2(t+1) \max_{0 \leq j \leq t} \left\{ \Pr\left[P_j(x) \in \mathbb{Z}_N \setminus \mathbb{Z}_N^* \text{ and } P_j(x') \in \mathbb{Z}_N^* : x, x' \xleftarrow{\$} \mathscr{C}\right]\right\}$$

Bounding the Probability of $\mathscr{F}_{\text{equal}}$. By construction of the simulator, $\mathscr{F}_{\text{equal}}$ occurs if there exist $P_i, P_j \sqsubseteq P$ such that

$$(P_i(x) = P_j(x) \text{ and } P_i(x') \neq P_j(x')) \text{ or } (P_i(x) \neq P_j(x) \text{ and } P_i(x') = P_j(x')). \quad (4.1)$$

Note that we want to consider the event $\mathscr{F}_{\text{equal}}$, conditioned on that event $\mathscr{F}_{\text{test}}$ did not occur. Therefore we may assume that there exists no straight line program $P_k \sqsubseteq P$ such that $P_k(x) = \perp$ and $P_k(x') \neq \perp$, or vice versa. This allows us to simplify our analysis slightly, since in this case (4.1) is equivalent to

$$(P_i(x) \equiv_N P_j(x) \text{ and } P_i(x') \not\equiv_N P_j(x')) \text{ or } (P_i(x) \equiv_N P_j(x) \text{ and } P_i(x') \not\equiv_N P_j(x')).$$

Thus, like in the previous section, we have

$$\Pr[\mathscr{F}_{\text{equal}} \mid \neg \mathscr{F}_{\text{test}}] \leq \sum_{-1 \leq i < j \leq t} 2 \cdot \Pr\left[P_i(x) \equiv_N P_j(x) \text{ and } P_i(x') \not\equiv_N P_j(x')\right]$$
$$\leq 2(t+2)(t+1) \max_{-1 \leq i < j \leq t} \left\{ \Pr\left[P_i(x) \equiv_N P_j(x) \text{ and } P_i(x') \not\equiv_N P_j(x')\right]\right\}$$
$$= 2(t^2 + 3t + 2) \max_{-1 \leq i < j \leq t} \left\{ \Pr\left[P_i(x) \equiv_N P_j(x) \text{ and } P_i(x') \not\equiv_N P_j(x')\right]\right\}$$

Bounding the Probability of $\mathscr{F}$. Summing up, we obtain that the total probability of $\mathscr{F}$ is at most

$$\Pr[\mathscr{F}] \leq \Pr[\mathscr{F}_{\text{test}}] + \Pr[\mathscr{F}_{\text{equal}}] \leq$$

$$2(t^2 + 3t + 2) \max_{-1 \leq i < j \leq t} \left\{ \Pr\left[P_i(x) \equiv P_j(x) \text{ and } P_i(x') \not\equiv P_j(x') : x, x' \xleftarrow{\$} \mathscr{C} \right] \right\}$$

$$+ 2(t+1) \max_{0 \leq k \leq t} \left\{ \Pr\left[P_k(x) \in \mathbb{Z}_N \setminus \mathbb{Z}_N^* \text{ and } P_k(x') \in \mathbb{Z}_N^* : x, x' \xleftarrow{\$} \mathscr{C} \right] \right\}.$$

4.6.3 Bounding the Success Probability

Since all computations of $\mathscr{A}$ are independent of the challenge value x in the simulation game, any algorithm has only the trivial success probability when interacting with the simulator. Thus the success probability of any algorithm when interacting with the original oracle is bounded by

$$\begin{aligned}
1/2 + \varepsilon &= \Pr[\text{Succ}_0(\mathscr{A})] = \Pr[\text{Succ}_1(\mathscr{A})] \\
&\leq \Pr[\text{Succ}_2(\mathscr{A})] + \Pr[\mathscr{F}] \\
&\leq 1/2 + \Pr[\mathscr{F}],
\end{aligned}$$

which implies

$$\varepsilon \leq \Pr[\mathscr{F}].$$

4.6.4 The Factoring Algorithm

Consider a factoring algorithm $\mathscr{B}$ which samples a random element $x \in \mathscr{C}$ and runs $\mathscr{A}$ as a subroutine by implementing the generic ring oracle for $\mathscr{A}$. That is, it performs all computations queried by $\mathscr{A}$ to $x \in \mathbb{Z}_N$.

In parallel, $\mathscr{B}$ applies all queried operations to $y \in \mathbb{Z}_N$, where $y \xleftarrow{\$} \mathscr{U}[\mathscr{C}]$ is chosen uniformly random at the beginning of the game. Moreover, each time a triple $(i, j, \circ)$ is appended to P, $\mathscr{B}$ computes

- $\gcd(P(y), N)$, and

- $\gcd(P(y) - P_i(y), N)$ for all $i \in \{-1, \ldots, |P| - 1\}$.

Running time. $\mathscr{B}$ samples random values $x \xleftarrow{\$} \mathscr{C}$ and $y \xleftarrow{\$} \mathscr{U}[\mathscr{C}]$. Since by assumption $\mathscr{A}$ submits t queries, $\mathscr{B}$ has to perform at most $2t$ operations in $\mathbb{Z}_N$ in order to perform all computations queried by $\mathscr{A}$ simultaneously on $x \in \mathscr{C}$ and $y \in \mathscr{U}[\mathscr{C}]$. In addition, $\mathscr{B}$ performs at most $(t+2)^2$ gcd-computations on $\lceil \log_2 N \rceil$-bit numbers.

Success probability. $\mathscr{B}$ evaluates any straight line program P_k with a uniformly random element y of $\mathscr{U}[\mathscr{C}]$. In particular, $\mathscr{B}$ computes $\gcd(P_k(y),N)$ for $y \overset{\$}{\leftarrow} \mathscr{U}[\mathscr{C}]$ and the straight line program $P_k \sqsubseteq P$ satisfying

$$\Pr\left[P_k(x) \in \mathbb{Z}_N \setminus \mathbb{Z}_N^* \text{ and } P_k(x') \in \mathbb{Z}_N^* : x,x' \overset{\$}{\leftarrow} \mathscr{C}\right]$$
$$= \max_{0 \le k \le t}\left\{\Pr\left[P_k(x) \in \mathbb{Z}_N \setminus \mathbb{Z}_N^* \text{ and } P_k(x') \in \mathbb{Z}_N^* : x,x' \overset{\$}{\leftarrow} \mathscr{C}\right]\right\}.$$

Let

$$\gamma_1 := \max_{0 \le k \le t}\{\Pr[P_k(x) \in \mathbb{Z}_N \setminus \mathbb{Z}_N^* \text{ and } P_k(x') \in \mathbb{Z}_N^* : x,x' \overset{\$}{\leftarrow} \mathscr{C}]\},$$

then by Lemma 4.3 algorithm $\mathscr{B}$ finds a factor in this step with probability at least γ_1.

Moreover, $\mathscr{B}$ evaluates any pair P_i, P_j of straight line programs in P with a uniformly random element $y \overset{\$}{\leftarrow} \mathscr{U}[\mathscr{C}]$ and computes $\gcd(P_i(y) - P_j(y), N)$. So in particular $\mathscr{B}$ computes $\gcd(P_i(y) - P_j(y), N)$ with $y \overset{\$}{\leftarrow} \mathscr{U}[\mathscr{C}]$ for the pair of straight line programs $P_i, P_j \sqsubseteq P$ satisfying

$$\Pr\left[P_i(x) \equiv_N P_j(x) \text{ and } P_i(x') \not\equiv_N P_j(x') : x,x' \overset{\$}{\leftarrow} \mathscr{C}\right]$$
$$= \max_{-1 \le i < j \le t}\left\{\Pr\left[P_i(x) \equiv_N P_j(x) \text{ and } P_i(x') \not\equiv_N P_j(x') : x,x' \overset{\$}{\leftarrow} \mathscr{C}\right]\right\}.$$

Let

$$\gamma_2 := \max_{-1 \le i < j \le t}\{\Pr[P_i(x) \equiv_N P_j(x) \text{ and } P_i(x') \not\equiv_N P_j(x') : x,x' \overset{\$}{\leftarrow} \mathscr{C}],\}$$

then by Lemma 4.4 algorithm $\mathscr{B}$ succeeds in this step with probability at least γ_2.

So, if we set $\gamma := \max\{\gamma_1, \gamma_2\}$, the total success probability of algorithm $\mathscr{B}$ is at least γ.

Relating the success probability of $\mathscr{B}$ to the advantage of $\mathscr{A}$. Using the above definitions of γ_1, γ_2, and γ, the fact that $\varepsilon \le \Pr[\mathscr{F}]$, and the derived bound on $\Pr[\mathscr{F}]$, we can obtain a lower bound on γ by

$$\varepsilon \le \Pr[\mathscr{F}] \le 2(t+1)\gamma_1 + 2(t^2 + 3t + 2)\gamma_2 \le 2(t^2 + 4t + 3)\gamma,$$

which implies the inequality

$$\gamma \ge \frac{\varepsilon}{2(t^2 + 4t + 3)}.$$

Therefore the success probability of $\mathscr{B}$ is at least

$$\frac{\varepsilon}{2(t^2+4t+3)}.$$

4.7 Applications

In this section, we apply our general theorem to two specific subset membership problems with high cryptographic relevance. The first application shows that computing Jacobi symbols modulo N with generic ring algorithms is as hard as factoring N. Since there exist efficient non-generic algorithms computing the Jacobi symbol, this shows that a proof in the generic ring model can not give any evidence towards the hardness of a computational problem.

Then we apply our main theorem to the well-known quadratic residuosity problem. It is unknown whether there exists an efficient algorithm for this problem, and it is widely conjectured that this problem is hard if factoring the modulus N is hard. We show that any algorithm solving this problem efficiently needs to exploit specific properties of the representation of elements of $\mathbb{Z}_N$ (possibly in a way similar to known algorithms for computing Jacobi symbols).

4.7.1 Computing the Jacobi Symbol with Generic Ring Algorithms

In order to define and analyze the *Jacobi symbol* we need the *Legendre symbol*. For an integer x and a prime p the Legendre symbol $(x \mid p)$ of x modulo p is defined as

$$(x \mid p) = \begin{cases} 0, & \text{if } \gcd(x,p) \neq 1, \\ 1, & \text{if } \gcd(x,p) = 1 \text{ and } x \text{ has a square root modulo } p, \\ -1, & \text{if } \gcd(x,p) = 1 \text{ and } x \text{ has no square root modulo } p. \end{cases}$$

The Jacobi symbol generalizes the Legendre symbol from prime to composite moduli. If $N = \prod_{i=1}^{l} p_i^{e_i}$ is the prime factor decomposition of N, then the Jacobi symbol $(x \mid N)$ of an integer x modulo N is defined as

$$(x \mid N) := \prod_{i=1}^{\ell} (x \mid p_i)^{e_i}, \tag{4.2}$$

where $(x \mid p_i)$ is the Legendre symbol. There exists an algorithm computing the Jacobi symbol $(x \mid N)$ efficiently, even if the factorization of N is not given, using the law of quadratic reciprocity. See [Sho08, Chapter 12.3], for instance.

Properties of the Jacobi symbol. In the sequel we will consider the problem of computing the Jacobi symbol as a subset membership problem over $\mathbb{Z}_N$. To this end, let us summarize some properties of the Jacobi symbol, which will become relevant.

1. Note that for $x \in \mathbb{Z}_N^*$ we have $(x \mid N) \in \{1, -1\}$. Let

$$J_N := \{x \in \mathbb{Z}_N^* : (x \mid N) = 1\}$$

   be the set of elements of $\mathbb{Z}_N$ having Jacobi symbol 1. Thus, we can perceive the problem of computing the Jacobi symbol as a subset membership problem $(\mathscr{C}, \mathscr{V})$ over $\mathbb{Z}_N$ with $\mathscr{C} = \mathbb{Z}_N^*$ and $\mathscr{V} = J_N$.

2. The cardinality $|J_N|$ of the set of elements having Jacobi symbol 1 depends on whether N is a square in $\mathbb{N}$. We have

$$|J_N| = \begin{cases} \varphi(N)/2, & \text{if } N \text{ is not a square in } \mathbb{N}, \\ \varphi(N), & \text{if } N \text{ is a square in } \mathbb{N}, \end{cases}$$

   where $\varphi(\cdot)$ is the Euler totient function [Sho08, Chapter 2.6]. This is an immediate consequence of the definition of the Jacobi symbol.

Now we are ready to apply our main theorem to show that there is no efficient *generic* ring algorithm computing the Jacobi symbol efficiently, unless factoring N is easy.

Theorem 4.2
Let $N = \prod_{i=1}^{\ell} p_i^{e_i}$. Suppose there exist a generic ring algorithm $\mathscr{A}$ (ε, t)-solving the subset membership problem given by $(\mathscr{C}, \mathscr{V})$ with $\mathscr{C} = \mathbb{Z}_N^*$ and $\mathscr{V} = J_N$. Then there exists an algorithm $\mathscr{B}$ finding a non-trivial factor of N with probability at least

$$\frac{\varepsilon}{2(t^2 + 4t + 3)}$$

by running $\mathscr{A}$ once, performing at most $2t$ additional operations in $\mathbb{Z}_N$ and at most $(t+2)^2$ gcd-computations on $\lceil \log_2 N \rceil$-bit numbers, and sampling two random elements from $\mathbb{Z}_N^*$.

PROOF. If n is a square in $\mathbb{N}$ then the theorem is trivially true, since in this case it is easy to find a factor of N. Therefore we only need to consider the case where N is not a square.

Note that in this case we have $2 \cdot |J_N| = \varphi(N) = |\mathbb{Z}_N^*| = |\mathscr{C}|$. Furthermore, it holds that $\mathscr{U}[\mathscr{C}] = \mathscr{U}[\mathbb{Z}_N^*] = \mathbb{Z}_N^* = \mathscr{C}$, which implies that $\mathscr{C}$ is homogeneous. The result follows by applying Theorem 4.1. $\qquad\square$

4.7.2 The Generic Quadratic Residuosity Problem and Factoring

Let us denote with $QR_N \subseteq \mathbb{Z}_N$ the set of *quadratic residues* modulo N, i.e.

$$QR_N := \{x \in \mathbb{Z}_N^* : x \equiv y^2 \bmod n, y \in \mathbb{Z}_N^*\}.$$

It holds that $QR_N \subseteq J_N$, and therefore given $x \in \mathbb{Z}_N \backslash J_N$ it is easy to decide that x is not a quadratic residue by computing the Jacobi symbol.

Definition 4.6
The *quadratic residuosity problem* [GM84] is the subset membership problem given by $\mathscr{C} = J_N$ and $\mathscr{V} = QR_N$.

If $N = pq$ is the product of two different odd primes, then it holds that

$$|QR_N| = \varphi(N)/4 \quad \text{and} \quad |J_N| = \varphi(N)/2,$$

see for instance [Sho08, p.348]. Thus, for $N = pq$ we have $2 \cdot |\mathscr{V}| = |\mathscr{C}|$.

Lemma 4.5
Let $N = p_1 p_2$ be the product of two different odd primes. Then J_N is homogeneous.

PROOF. We have to show that for each $c_1 \in \mathbb{Z}_{p_1}$ and $c_2 \in \mathbb{Z}_{p_2}$ holds that

$$\Pr[x \equiv c_1 \bmod p_1 : x \xleftarrow{\$} J_N] = \Pr[x \equiv c_1 \bmod p_1 : x \xleftarrow{\$} \mathbb{Z}_N^*] \qquad (4.3)$$

$$\text{and } \Pr[x \equiv c_2 \bmod p_2 : x \xleftarrow{\$} J_N] = \Pr[x \equiv c_2 \bmod p_2 : x \xleftarrow{\$} \mathbb{Z}_N^*]. \qquad (4.4)$$

In the sequel we will consider case (4.3), case (4.4) is identical.

Note first that we have $\Pr[x \equiv 0 \bmod p_1 : x \xleftarrow{\$} \mathbb{Z}_N^*] = 0$, and for each $c_1 \in \mathbb{Z}_{p_1}^*$ with $c_1 \not\equiv 0 \bmod p_1$ we have

$$\Pr[x \equiv c_1 \bmod p_1 : x \xleftarrow{\$} \mathbb{Z}_N^*] = 1/(p_1 - 1).$$

Since $J_N \subseteq \mathbb{Z}_N^*$ we have $\Pr[x \equiv 0 \bmod p_1 : x \xleftarrow{\$} J_N] = 0$, thus it only remains to show that

$$\Pr[x \equiv c_1 \bmod p_1 : x \xleftarrow{\$} J_N] = 1/(p_1 - 1)$$

holds for all for all $c_1 \not\equiv 0 \bmod p_1$.

Let ψ denote the isomorphism $\psi : \mathbb{Z}_{p_1} \times \mathbb{Z}_{p_2} \to \mathbb{Z}_N$. Then the set J_N consists of all elements $x = \psi(x_1, x_2) \in \mathbb{Z}_N^*$ such that $(x_1 \mid p_1) \cdot (x_2 \mid p_2) = 1$, which is equivalent to $(x_1 \mid p_1) = (x_2 \mid p_2)$. Thus we have

$$
\begin{aligned}
J_N &= \{ x \in \mathbb{Z}_N^* : (x \mid n) = 1 \} \\
&= \{ \psi(x_1, x_2) \in \mathbb{Z}_N^* : (x_1 \mid p_1) \cdot (x_2 \mid p_2) = 1 \} \\
&= \{ \psi(x_1, x_2) \in \mathbb{Z}_N^* : (x_1 \mid p_1) = (x_2 \mid p_2) \}
\end{aligned}
$$

It is well-known that for each odd prime p_2 holds that

$$
\left| \{ x_2 \in \mathbb{Z}_{p_2}^* : (x_2 \mid p_2) = 1 \} \right| = \left| \{ x_2 \in \mathbb{Z}_{p_2}^* : (x_2 \mid p_2) = -1 \} \right| = (p_2 - 1)/2.
$$

Therefore for each element $x_1 \in \mathbb{Z}_{p_1}^*$ there are exactly $(p_2 - 1)/2$ elements $x_2 \in \mathbb{Z}_q^*$ such that $(x_1 \mid p_1) \cdot (x_2 \mid p_2) = 1$, and thus $\psi(x_1, x_2) \in J_N$. Thus we have

$$
\left| \{ x \in J_N : x \equiv c_1 \bmod p_1 \} \right| = (p_2 - 1)/2.
$$

This yields that for each $c_1 \in \mathbb{Z}_{p_1}^*$ with $c_1 \not\equiv 0 \bmod p_1$ we have

$$
\begin{aligned}
\Pr[x \equiv c_1 \bmod p_1 : x \xleftarrow{\$} J_N] &= \frac{\left| \{ x \in J_N : x \equiv c_1 \bmod p_1 \} \right|}{|J_N|} \\
&= \frac{(p_2 - 1)/2}{(p_1 - 1)(p_2 - 1)/2} = \frac{1}{p_1 - 1}.
\end{aligned}
$$

$\square$

Given the factorization of an integer N, the quadratic residuosity problem in $\mathbb{Z}_N$ can be solved easily by a generic ring algorithm. Thus, in order to show the equivalence of generic quadratic residuosity and factoring, we have to prove the following theorem.

Theorem 4.3
Let $N = pq$ be the product of two different odd primes. Suppose there exist a generic ring algorithm $\mathscr{A}$ (ε, t)-solving the subset membership problem given by $(\mathscr{C}, \mathscr{V})$ with $\mathscr{C} = J_N$ and $\mathscr{V} = \mathrm{QR}_N$. Then there exists an algorithm $\mathscr{B}$ finding a non-trivial factor of N with probability at least

$$
\frac{\varepsilon}{2(t^2 + 4t + 3)}
$$

by running $\mathscr{A}$ once, performing at most $2t$ additional operations in $\mathbb{Z}_N$ and at most $(t + 2)^2$ gcd-computations on $\lceil \log_2 Nn \rceil$-bit numbers, and sampling each one random element from J_N and $\mathbb{Z}_N^*$.

PROOF. If $N = pq$ is the product of two different odd primes, then we have $\mathcal{U}[\mathscr{C}] = \mathcal{U}[J_N] = \mathbb{Z}_N^*$ and $|\mathscr{C}| = |J_N| = 2 \cdot |\mathrm{QR}_N| = 2 \cdot |\mathcal{V}|$. The result follows by applying Lemma 4.5 and Theorem 4.1. □

To show that $\mathscr{B}$ factors N efficiently, it remains to show that $\mathscr{B}$ can efficiently sample uniformly random elements of J_N. Consider an algorithm $\mathscr{B}$ which samples uniformly random elements x from $\mathbb{Z}_N$ until $x \in J_N$ (note that $\mathscr{B}$ can test efficiently whether $x \in J_N$ by running the algorithm from [Sho08, Chapter 12.3]). Moreover, for $x \overset{\$}{\leftarrow} \mathbb{Z}_N$ and large N, we have

$$\Pr[x \in J_N] = \frac{|J_N|}{|\mathbb{Z}_N|} = \frac{\varphi(N)/2}{N} \approx \frac{1}{2}$$

thus we may expect that $\mathscr{B}$ finds a suitable x very quickly.

4.7.3 Analysis of Search Problems

In the proof of Theorem 4.1 we have constructed a simulator for a generic ring oracle for the ring $\mathbb{Z}_N$. When interacting with the simulator, all computations are independent of the secret challenge value x. Therefore we have been able to conclude that any generic algorithm has only the trivial probability of success in solving certain decisional problems (namely the considered subset membership problems) when interacting with the simulator. Moreover, we have shown that any algorithm distinguishing between simulator and original oracle can be turned into a factoring algorithm with (asymptotically) the same running time.

In contrast to *decisional* problems, where the algorithm outputs a bit, our construction of the simulator can also be applied to prove the generic hardness of *search* problems where the algorithm outputs a ring element or integer. Let us sketch two possibilities. The first one is to formulate a suitable subset membership problem which reduces to the considered search problem and then apply Theorem 4.1. Another possibility is to use our construction of the simulator to bound the probability of a simulation failure relative to factoring. In order to bound the success probability in the simulation game, it remains to show that there exists no *straight line program* solving the considered problem efficiently under the factoring assumption. This implies the following theorem.

Theorem 4.4

Let $N \in \mathbb{N}$ such that finding a factor of N is hard. For any search problem in $\mathbb{Z}_N$ with uniformly random challenge $x \overset{\$}{\leftarrow} \mathbb{Z}_N$, there exist a generic ring algorithm $\mathscr{A}$ solving the problem efficiently *if and only if* there exists a straight line program solving the problem efficiently.

5 The Generic Composite Residuosity Problem

The decisional composite residuosity problem (DCR) was introduced by Paillier in [Pai99]. Essentially, the problem is to distinguish an N-th residue $r^N \bmod N^2$, where $r \xleftarrow{\$} \mathbb{Z}_N^*$, from a random element of $\mathbb{Z}_{N^2}^*$, where $N = PQ$ is a RSA modulus. This problem was later generalized by Damgård and Jurik [DJ01] and Catalano et al. [CGHGN01].

It is known that the DCR problem can be reduced to factoring N. However, no reduction from a well-studied computational problem to the DCR problem is known so far, neither in the standard model nor in an idealized model of computation. Since the assumption that solving this problem is hard has found many interesting cryptographic applications, such as efficient instantiations of additively homomorphic encryption [Pai99], lossy trapdoor functions [RS08, FGK⁺10], public-key encryption schemes secure against selective-opening attacks [HLOV09], and many more, it is interesting to study the validity of this assumption.

We analyze a generalized version of the DCR problem, which includes the variants from [Pai99, DJ01, CGHGN01] as special cases, in the generic ring model. As illustrated in Section 5.4, the general theorem from Chapter 4, stating the generic hardness of a large class of decisional problems, can not be applied to the DCR problem. Therefore we devise a different argument that allows us to relate the generic DCR problem modulo N^ℓ to the so-called Hensel-RSA problem [CNS02]. Reductions to the Hensel-RSA problem are known from the well-studied RSA problem [RSA78], and from the composite residuosity class problem [CNS02], depending on the considered algebraic setting.

5.1 Related Work

Decisional Composite Residuosity and Variants. Let $N = PQ$ be a RSA modulus, and let $y \xleftarrow{\$} \mathbb{Z}_{N^2}^*$ and $b \xleftarrow{\$} \{0,1\}$. The DCR problem, originally introduced in [Pai99], is to compute b on input $x = y^{N^b} \bmod N^2$. Thus, the DCR problem asks to distinguish an N-th residue modulo N^2 from a random element of $\mathbb{Z}_{N^2}^*$. This problem was generalized by Damgård and Jurik [DJ01] to computing b on input

$x = y^{N^{b(\ell-1)}} \bmod N^\ell$ for $\ell \geq 2$. In [CGHGN01] a variant was introduced, where instead of N-th residues e-th residues are considered with $\gcd(e, \phi(N)) = 1$.

Hensel-RSA. We are going to relate the generic DCR problem to the so-called *Hensel-RSA problem*. This problem was introduced and studied by Catalano, Nguyen, and Stern in [CNS02]. The (N, ℓ, e)-Hensel-RSA problem is to compute $x^e \bmod N^\ell$ on input $x^e \bmod N$ for random $x \xleftarrow{\$} \mathbb{Z}_N$. It was shown in [CNS02] that the Hensel-RSA is equivalent to the classical RSA problem [RSA78] if $\ell \geq 3$, and harder than the composite residuosity class problem [CNS02] for $\ell \geq 2$ (see Section 5.5 for details).

5.2 Results of This Chapter

We define a generalized DCR problem, which captures the variants from [Pai99, DJ01, CGHGN01] as special cases, and describe it in the generic ring model. We give an intuitive explanation why the general theorem on the hardness of subset membership problems from Chapter 4 is not applicable to DCR-type problems. Our main result is then a proof that solving the generalized DCR problem generically is equivalent to solving the Hensel-RSA problem. In combination with the work of [CNS02], this implies that solving DCR-type problems generically is hard, if the RSA problem or the composite residuosity class problem are hard.

5.3 Generic Decisional Composite Residuosity

The following generalized DCR problem captures the variants of [Pai99, DJ01, CGHGN01] as special cases.

Definition 5.1
We say the that the (N, ℓ, e)-*DCR problem* is $(\varepsilon_{\mathsf{DCR}}, t)$-hard, if

$$\left| \Pr[\mathscr{A}(N, \ell, e, x_0) = 1] - \Pr[\mathscr{A}(N, \ell, e, x_1) = 1] \right| \leq \varepsilon_{\mathsf{DCR}},$$

where $x_0 \xleftarrow{\$} \mathbb{Z}_{N^\ell}^*$ and $x_1 := x_0^e \bmod N^\ell$, for all algorithms $\mathscr{A}$ running in time t.

Then the classical DCR problem from [Pai99] is called the $(N, 2, N)$-DCR problem, the $(N, \ell, N^{\ell-1})$-DCR problem with $\ell \geq 2$ is due to Damgård and Jurik [DJ01], and the $(N, 2, e)$-DCR problem with $\gcd(e, \phi(N)) = 1$ is from [CGHGN01].

Generic Decisional Composite Redisuosity. To analyze the (N, ℓ, e)-DCR problem in the generic ring model, we consider algorithms interacting with an oracle

$$\mathcal{O}(\mathbb{Z}_{N^\ell}, \Pi, \Sigma; \vec{x}),$$

where $\vec{x} = (1, x)$ for $x = y^{e^b} \bmod N^\ell$ with $y \xleftarrow{\$} \mathbb{Z}_{N^\ell}^*$ and $b \xleftarrow{\$} \{0, 1\}$, $\Pi = \{+, -, \cdot\}$ and $\Sigma = \{=\}$. Note that we are using the generic ring model of [LR06], where no explicit division operation $\div$ is included. This is unfortunately necessary, due to the fact that the proving technique applied below allows to consider only this class of generic ring algorithms.

Definition 5.2
We say the that the generic (N, ℓ, e)-decisional composite residuosity problem is $(\varepsilon_{\mathsf{DCR}}, t)$-hard, if

$$\left| \Pr[\mathcal{A}^{\mathcal{O}}(N, \ell, e) = 1 \mid b = 0] - \Pr[\mathcal{A}^{\mathcal{O}}(N, \ell, e) = 1 \mid b = 1] \right| \leq \varepsilon_{\mathsf{DCR}}.$$

for all algorithms $\mathcal{A}$ running in time t.

5.4 Why Theorem 4.1 is Not Applicable

In Chapter 4 we have proven a general theorem on the hardness of a large class of subset membership problems in $\mathbb{Z}_N$. Unfortunately, this theorem is not applicable to the DCR problem, and it seems hard to generalize it to the DCR case, as we illustrate in the sequel.

Limitations of Theorem 4.1 from Chapter 4. Recall that Theorem 4.1 requries that challenges are sampled *uniformly random* from $\mathscr{C}$. This requirement seems to be inherent in the argument applied in the proof, since the factoring algorithm $\mathscr{B}$ works only in this case. We do not know how to prove the same statement for subset membership problems where this does not hold.

Applicability to Paillier's DCR problem. The DCR problem [Pai99] is a subset membership problem which does not have the required property. In the notation from Chapter 4, Paillier's DCR problem is the subset membership problem $(\mathscr{C}, \mathscr{V})$ with

$$\mathscr{C} = \mathbb{Z}_{N^2}^* \quad \text{and} \quad \mathscr{V} = \{r^N \bmod N^2 : r \in \mathbb{Z}_{N^2}^*\}.$$

Here we have $|\mathscr{C}| = \phi(N) \cdot N$ and $|\mathscr{V}| = \phi(N)$. The probability that a random element of $\mathscr{C}$ is an element of $\mathscr{V}$ is

$$\Pr[x \in \mathscr{V} : x \xleftarrow{\$} \mathscr{C}] = 1/N,$$

and thus negligibly small. Therefore sampling a uniformly random challenge $x \xleftarrow{\$}$ $\mathscr{C}$ would nearly always yield an N-th non-residue.

In the sequel we thus have to develop another technique to study the generic hardness of Paillier's DCR problem and its variants.

5.5 Hardness of Hensel-RSA Lifting

We are going to relate the generic hardness of DCR to solving the so-called *Hensel-RSA* problem introduced by Catalano, Nguyen, and Stern [CNS02].

Definition 5.3
We say that the (N, ℓ, e)-Hensel-RSA problem is $(\varepsilon_{\text{HRSA}}, t)$-hard, if

$$\Pr[\mathscr{A}(N, \ell, e, x^e \bmod N) = x^e \bmod N^\ell] \leq \varepsilon_{\text{HRSA}}$$

for all algorithms $\mathscr{A}$ running in time t.

As shown by Catalano et al. in [CNS02], the following two problems reduce to solving Hensel-RSA.

- RSA PROBLEM. Let e be an integer such that $\gcd(e, \phi(N)) = 1$, where ϕ denotes Euler's Phi function. The (N, e)-RSA problem is to compute $x \in \mathbb{Z}_N$ on input $x^e \bmod N$, where $x \xleftarrow{\$} \mathbb{Z}_N$ is chosen uniformly random. A special case is the (N, N)-RSA problem, where the exponent e equals the modulus N.

- COMPOSITE RESIDUOSITY CLASS PROBLEM. The (N, g)-composite residuosity class problem is: given a random element $y \xleftarrow{\$} \mathbb{Z}_{N^2}^*$, compute $m \in \mathbb{Z}_N$ such that

$$y = g^m r^N \bmod N^2. \tag{5.1}$$

In [Pai99] it was proven that for each $y \in \mathbb{Z}_{N^2}^*$ there exists a *unique* pair $(m, r) \in \mathbb{Z}_N \times \mathbb{Z}_N^*$ satisfying (5.1). Note that the composite residuosity class problem is exactly the problem of decrypting a Paillier ciphertext.

Both above problems are assumed to be hard if factoring N is hard. Strong evidence towards this assumption was given by Catalano et al. [CNS02], who proved the following lemma.

Lemma 5.1 ([CNS02])
Let $N = PQ$ be a RSA modulus.

- If there exists an algorithm solving the (N, ℓ, e)-Hensel-RSA problem for $\ell \geq 3$, then there exists an algorithm solving the (N, e)-RSA problem.
- If there exists an algorithm solving the $(N, 2, N)$-Hensel-RSA problem, then there exists an algorithm solving the (N, g)-composite residuosity class problem.

5.6 Analysis of the Generic DCR Problem

We relate the hardness of solving the generic DCR problem to solving the Hensel-RSA problem. In combination with the above results of Catalano et al., this relates the generic DCR problem to the RSA problem if $\ell \geq 3$, and to the composite residuosity class problem if $\ell \geq 2$.

Theorem 5.1
If solving the (N, ℓ, e)-Hensel-RSA problem is $(\varepsilon_{\mathsf{HRSA}}, t)$-hard, then solving the generic (N, ℓ, e)-decisional composite residuosity problem over $\mathbb{Z}_{N^\ell}$ is (ε', t') hard with

$$\varepsilon' \leq \frac{q^2}{\varepsilon_{\mathsf{HRSA}}} + \frac{q^2}{P} + \frac{2N^{\ell-1}(P+Q+1)}{N^\ell} \quad \text{and} \quad t' + q \approx t,$$

where P is the smallest prime factor of N and q is an upper bound on the number of oracle queries issued by the generic ring algorithm $\mathscr{A}$.

PROOF. We proceed in a sequence of games. We start with a game where the algorithm interacts with an oracle with inital state $(1, x)$ where $x \xleftarrow{\$} \mathbb{Z}_{N^\ell}^*$ is a random element of $\mathbb{Z}_{N^\ell}^*$, and we end up with a game where the algorithm interacts with an oracle where $x \equiv y^e \bmod N^\ell$ is an e-th residue. We will obtain the result by bounding the probability that the algorithm distinguishes Game i from Game $i-1$ for all i. In the following let $\mathscr{O}_i$ denote the oracle that $\mathscr{A}$ interacts with in Game i.

Game 0. This game corresponds to the generic DCR experiment described above, with $b = 0$. That is, the algorithm interacts with an oracle $\mathscr{O}_0$ whose initial list contents is $L_1 = (1, x)$, where $x \xleftarrow{\$} \mathbb{Z}_{N^\ell}^*$ is a random element of $\mathbb{Z}_{N^\ell}^*$. We have

$$\Pr[\mathscr{A}^{\mathscr{O}_0}(N, \ell, e) = 1] = \Pr[\mathscr{A}^{\mathscr{O}}(N, \ell, e) = 1 \mid b = 0].$$

Game 1. We change the way the challenge x is sampled. Instead of choosing $x \xleftarrow{\$} \mathbb{Z}_{N^\ell}^*$, $\mathcal{O}_1$ samples $x \xleftarrow{\$} \mathbb{Z}_{N^\ell}$. We assume that $\mathcal{O}_1$ does so by choosing two integers $x_0 \xleftarrow{\$} \mathbb{Z}_N$ and $x_1 \xleftarrow{\$} \mathbb{Z}_{N^{\ell-1}}$ and setting $x = x_1 N + x_0$. This is equivalent to sampling $x \xleftarrow{\$} \mathbb{Z}_{N^\ell}$. Otherwise $\mathcal{O}_1$ proceeds exactly like $\mathcal{O}_0$.

We have $N^\ell - \phi(N^\ell) = N^\ell - N^{\ell-1}(P-1)(Q-1) = N^{\ell-1}(P+Q+1)$, and thus

$$\left| \Pr[\mathscr{A}^{\mathcal{O}_1}(N,\ell,e) = 1] - \Pr[\mathscr{A}^{\mathcal{O}_0}(N,\ell,e) = 1] \right| \leq \frac{N^{\ell-1}(P+Q+1)}{N^\ell}.$$

Game 2. In this game we modify the way the integer x is sampled. Oracle $\mathcal{O}_2$ samples $x_0 \xleftarrow{\$} \mathbb{Z}_N$ and $x_1 \xleftarrow{\$} \mathbb{Z}_{N^{\ell-1}}$, but instead of performing all computations on integers, oracle $\mathcal{O}_2$ uses polynomials from $\mathbb{Z}_{N^\ell}[X]$ for the internal representation of ring elements. To this end, it proceeds as follows.

1. The list L is initialized with $L_1 = 1$ and $L_2 = x = XN + x_0$. Note that the variable X is used instead of x_1 (x_1 is not used throughout the game, but it is useful to have it defined in order to compare Game 2 to Game 1 in the analysis below).

2. Whenever the algorithm asks to perform a computation $\circ \in \{+,-,\cdot\}$ on two list elements L_i, L_j, the oracle computes

$$L_k = L_i \circ L_j.$$

   Note that each list element L_i can be written as a polynomial $L_i(X) = (a_i X + b_i)N + c_i$, where $a_i, b_i \in \mathbb{Z}_{N^{\ell-1}}$ and $c_i \in \mathbb{Z}_N$.

3. Whenever the algorithm asks to perform an equality test on two list elements L_i, L_j, then $\mathcal{O}_2$ returns 1 if

$$(a_i, b_i, c_i) = (a_j, b_j, c_j),$$

   and 0 otherwise.

Observe that $\mathcal{O}_2$ simulates $\mathcal{O}_1$ perfectly, unless $\mathcal{O}_2$ replies with 0 on an equality test query where $\mathcal{O}_1$ would have returned 1 (the opposite case is impossible). Note that this happens only if

$$(a_i, b_i, c_i) \neq (a_j, b_j, c_j) \qquad \text{but} \qquad L_i(x_1) \equiv L_j(x_1) \bmod N^\ell.$$

Since $c_i \neq c_j$ implies $L_i(x_1) \not\equiv L_j(x_1) \bmod N^\ell$, it suffices to consider the case where $c_i = c_j$ and $(a_i, b_i) \neq (a_j, b_j)$. In this case we have

$$a_i x_1 + b_i \equiv a_j x_1 + b_j \bmod N^{\ell-1},$$

or equivalently

$$(a_i - a_j)x_1 + (b_i - b_j) \equiv 0 \bmod N^{\ell-1}, \tag{5.2}$$

where x_1 is uniformly random and independent of the algorithm's view. If P denotes the smallest prime factor of N, then the polynomial (5.2) of degree one has at most one root modulo P. Thus, the probability that by issuing q oracle queries the algorithm computes two pairs (a_i, b_i) and (a_j, b_j) such that $\mathcal{O}_2$ fails to simulate $\mathcal{O}_1$ is at most q^2/P. This implies

$$\left| \Pr[\mathscr{A}^{\mathcal{O}_2}(N, \ell, e) = 1] - \Pr[\mathscr{A}^{\mathcal{O}_1}(N, \ell, e) = 1] \right| \leq \frac{q^2}{P}.$$

Game 3. This game corresponds to the generic DCR experiment described above, with $b = 1$. That is, we replace the simulator from the previous game with an oracle $\mathcal{O}_3$ whose initial list contents is $L_1 = (1, x)$. Here x is sampled by $\mathcal{O}_3$ by choosing $y \xleftarrow{\$} \mathbb{Z}_{N^\ell}$ and computing $x := y^N \bmod N^\ell$. We write x as $x = x_1 N + x_0$.

Note that $\mathcal{O}_2$ simulates $\mathcal{O}_3$ perfectly, unless

$$(a_i, b_i, c_i) \neq (a_j, b_j, c_j) \qquad \text{but} \qquad L_i(x_1) \equiv L_j(x_1) \bmod N^\ell.$$

Note that in this case the algorithm must have computed two pairs (a_i, b_i) and $(a_j, b_j))$ such that

$$a_i x_1 + b_i \equiv a_j x_1 + b_j \bmod N^{\ell-1},$$

or equivalently

$$x_1 \equiv (b_j - b_i)(a_i - a_j)^{-1} \bmod N^{\ell-1}. \tag{5.3}$$

Suppose there exists an algorithm $\mathscr{A}$ performing a sequence of at most q operations such that the probability that there exist two pairs (a_i, b_i) and $(a_j, b_j))$ such that (5.3) holds is at least ε. Then we have

$$\left| \Pr[\mathscr{A}^{\mathcal{O}_3}(N, \ell, e) = 1] - \Pr[\mathscr{A}^{\mathcal{O}_3}(N, \ell, e) = 1] \right| \leq \varepsilon.$$

We can construct an algorithm $\mathscr{B}$ solving the (N, ℓ, e)-Hensel-RSA problem as follows. $\mathscr{B}$ receives as input $x_0 = x^e \bmod N$ for random $x \xleftarrow{\$} \mathbb{Z}_N$. Note that x_0 is uniformly distributed over $\mathbb{Z}_N$, since the map $x \mapsto x^e \bmod N$ is a permutation over

$\mathbb{Z}_N$ if $\gcd(e, \phi(N)) = 1$. $\mathscr{B}$ runs $\mathscr{A}$ as a subroutine by implementing the simulator from Game 2 for $\mathscr{A}$. When $\mathscr{A}$ terminates, or after at most q oracle queries, $\mathscr{B}$ guesses two random indices $i, j \xleftarrow{\$} \{1, \ldots, q\}$, and computes and returns

$$\tau \equiv (b_j - b_i)(a_i - a_j)^{-1} \bmod N^{\ell-1}.$$

By assumption, with probability at least ε there exist two pairs (a_i, b_i) and (a_j, b_j)) such that (5.3) holds. With probability $1/q^2$, $\mathscr{B}$ guesses the indices i, j correctly, such that it obtains $\tau = x_1$. Thus we have $\varepsilon \leq q^2/\varepsilon_{\mathsf{HRSA}}$, and therefore

$$\left| \Pr[\mathscr{A}^{\mathscr{O}_3}(N, \ell, e) = 1] - \Pr[\mathscr{A}^{\mathscr{O}_2}(N, \ell, e) = 1] \right| \leq \frac{q^2}{\varepsilon_{\mathsf{HRSA}}}.$$

Game 4. We change the way the challenge is sampled. Instead of choosing $y \xleftarrow{\$} \mathbb{Z}_{N^\ell}$, $\mathscr{O}_4$ chooses $y \xleftarrow{\$} \mathbb{Z}_{N^\ell}^*$ and then computes $x := y^N \bmod N^\ell$. Thus we have

$$\Pr[\mathscr{A}^{\mathscr{O}_4}(N, \ell, e) = 1] = \Pr[\mathscr{A}^{\mathscr{O}}(N, \ell, e) = 1 \mid b = 1].$$

and, like in Game 1,

$$\left| \Pr[\mathscr{A}^{\mathscr{O}_4}(N, \ell, e) = 1] - \Pr[\mathscr{A}^{\mathscr{O}_3}(N, \ell, e) = 1] \right| \leq \frac{N^{\ell-1}(P + Q + 1)}{N^\ell}.$$

Collecting probabilities from Game 0 to Game 4 yields

$$\left| \Pr[\mathscr{A}^{\mathscr{O}}(N, \ell, e) = 1 \mid b = 0] - \Pr[\mathscr{A}^{\mathscr{O}}(N, \ell, e) = 1 \mid b = 1] \right|$$

$$\leq \frac{q^2}{\varepsilon_{\mathsf{HRSA}}} + \frac{q^2}{P} + \frac{2N^{\ell-1}(P + Q + 1)}{N^\ell}$$

$$\square$$

Generic disclaimer. We note again that it seems dangerous to perceive hardness results in the generic model as evidence towards a hardness assumption, since the result holds only for a (quite general, but still) restricted class of algorithms. This holds especially given the results from Chapter 4, showing that there exist *practical* (i.e. not contrived) computational problems which are provably hard to solve generically, but easy to solve in general.

However, we think that a proof in the generic model is still interesting from a cryptanalytic point of view, since it rules out a large class of algorithms (namely all algorithms that do not exploit specific properties of the representation of ring elements) solving the considered problem efficiently.

6 Semi-Generic Groups and Their Applications

The generic group model (GGM) is used frequently to provide evidence towards newly introduced hardness assumptions. In particular in the area of pairing-based cryptography numerous novel assumptions have been introduced over the last decade. Unfortunately, the GGM does not reflect many known properties of bilinear group settings. Not at least *currently known* algorithms for solving computational problems over bilinear groups are captured, and thus hardness results in this model are of limited significance.

In this chapter, we propose a novel black-box model, called the *semi-generic group model*, that is closer to the standard model and allows to make more meaningful security statements. We describe several instantiations of this model, which apply to both single-group settings and different types of group settings from pairing-based cryptography.

An inportant aspect of these models is that *the best algorithms currently known for solving algebraic problems commonly used in cryptography are semi-generic in nature*, and thus captured by the model. We demonstrate the usefulness of our new model by applying it exemplarily to study important assumptions, namely the computational Diffie-Hellman problem, the Co-Diffie-Hellman problem, and the bilinear decisional Diffie-Hellman problem. The presented techniques are rather general and can be adopted to study further hardness assumptions.

6.1 Motivation and Related Work

It is widely known that one has to take care when interpreting a proof in the generic group model as evidence towards the validity of a cryptographic hardness assumption or the security of a cryptosystem [Fis00, Den02, KM07], since it abstracts away from potentially many properties an adversary might be able to exploit in the real world. On the one hand, there exist cryptographic groups (such as certain elliptic curve groups) for which not many properties beyond the axioms of an algebraic group are known. Hence, modeling such groups as generic can be seen as a reasonable abstraction. On the other hand, there are groups featuring many further properties, which clearly makes the generic group model an inappropriate reflection for them. A prime example are multiplicative groups of finite fields or rings. These structures offer many well-understood properties beyond the group axioms, such as additional efficient algebraic operations (e.g., addition in the field or ring), and other properties of the group representation (e.g., the notion of prime integers and irreducible polynomials), that are simply ignored by the generic group model, but give rise to more efficient algorithms for certain problems (e.g., index calculus algorithms for computing discrete logarithms). *But should a minimal requirement on such an idealized model of computation not be that at least all currently known algorithms are captured?*

There exist some first approaches in the cryptographic literature to tackle this issue: The pseudo-free group model proposed by Hohenberger [Hoh03] and Rivest [Riv04] (see also [Mic05, CFW11]) does not treat a group as a black-box. Unfortunately, the definition of pseudo-freeness is very restrictive in the sense that a number of important groups (like all known-order groups) are immediately excluded and important problems, such as Diffie-Hellman-type problems, seem not to be covered by the model. Other approaches due to Leander and Rupp [LR06] and Aggarwal and Maurer [AM09] take into account that the RSA group $\mathbb{Z}_N^*$ is embedded in the ring $\mathbb{Z}_N$. They use the generic ring model (see Section 2.3), where an algorithm may perform both multiplication and addition operations in $\mathbb{Z}_N$ to show that breaking RSA is equivalent to factoring. Unfortunately, our work presented in Chapter 4 and [JS09] shows that even computing the *Jacobi symbol* [Sho08] is equivalent to factoring in this model. Thus this approach has not led to a satisfying abstraction of reality yet.

6.2 Results of This Chapter

We describe a novel black-box model of computation, which we call the *semi-generic group model*, and describe several instantiations of this model that apply both to single-group settings and to different types of settings from pairing-based cryptography. In contrast to the classical generic group model, the new model captures the best currently known algorithms for solving various algebraic problems commonly used in cryptography.

While the new model results from a simple technical modification to the classical generic group model, we need to develop new proof techniques which differ from classical proofs in the GGM. These new techniques rely on reduction techniques known from proofs in the standard model, since the (less natural) technique of deriving exponential complexity lower bounds is not applicable anymore.

As an exemplary application of the model, we analyze some important computational and decisional hardness assumptions from classical and pairing-based cryptography in our new model. This includes the Diffie-Hellman, Co-Diffie-Hellman, and decisional bilinear Diffie-Hellman problems.

The semi-generic group model should not be seen as a replacement for the careful analysis of cryptographic hardness assumptions, or reductions from thoroughly analyzed assumptions, in the standard model. It should rather be seen as a tool to get the currently best possible immediate evidence towards a hardness assumption, or to analyze the security of a practical cryptosystem that cannot be proven secure in the standard model.

6.3 An Extended Black-Box Model of Computation

In Chapter 2 we have described a general black-box model of computation. Two concrete instantiations of this general model, the generic group model (Section 2.2) and the generic bilinear group model (Section 2.4), have been used in various previous works to analyze both classical [Sho97, MW98, MW99, Mau05] and newly introduced cryptographic hardness assumptions in bilinear groups [KSW08, RLB⁺08, Boy08, BB08]. In this section, we will describe an extension of this model, and use it to instantiate what we call *semi-generic groups* and *semi-generic bilinear groups*.

The modification is technically rather simple. Recall that in Section 2.1 we have characterized a black-box oracle by $q+2$ sets $S_1, \ldots, S_q, \Pi, \Sigma$. We extend this characterization by another set $\Theta \subseteq \{1, \ldots, q\}$. Based on this additonal set, we allow a further type of queries an algorithm may ask to the oracle. In addition to computation queries from Π and relation queries from Σ, an algorithm in the semi-generic group model may ask *reveal*-queries of elements stored in the lists L_i, $i \in \Theta$. To this end, the algorithm submits two indices (i, j). The oracle responds as follows.

- If $i \notin \Theta$, then the oracle returns an error symbol $\bot$.

- If $i \in \Theta$, then the oracle returns the contents of $L_{i,j}$.

Thus, by making reveal-queries, an algorithm may learn the elements stored in the list L_i in their *standard* representation, if $i \in \Theta$.

6.4 Semi-Generic Groups

Let $\mathbb{G} = (S, \circ)$ be a group. In the *semi-generic group model* (sGGM) an algorithm interacts with an oracle

$$\mathcal{O}(S, \Pi, \Sigma, \Theta; \vec{x}),$$

where $\Pi = \{\circ\}$, $\Sigma = \{=\}$, and $\Theta = \{1\}$. As described in Section 2.2, $=$ denotes the (binary) equality relation, and the vector $\vec{x}$ is chosen according to the considered computational problem, and contains usually at least a generator of the group $\mathbb{G}$.

6.4.1 Relation to the Generic and Standard Model

It is obvious that the semi-generic group model considers a larger class of algorithms than the generic group model from Section 2.2, since it allows to exploit specific properties of the given representation of group elements. There are many trivial examples for computational problems which are provably "hard" in the black-box model, but "easy" to solve semi-generically. Consider, for instance, the discrete logarithm problem in the additive group $(\mathbb{Z}_p, +)$ of integers modulo p. However, the relation between the semi-generic group model and the standard model is not that obvious.

One may be tempted to think that the semi-generic group model is *equivalent* to the standard model: an algorithm may simply issue reveal-queries to all initial list elements in $L_1, \ldots, L_q$, and then perform all computations "offline", that is, independent of the oracle. If this was true, then the semi-generic group model would clearly not be useful. However, whether this is true or not depends on the considered computational problem. We explain this by giving two examples.

Example 6.1

Let $\mathbb{G} = (S, \circ)$ be a group of order p with generator g. Let $a, b, c \overset{s}{\leftarrow} \mathbb{Z}_p$ be uniformly random. We say that an algorithm $\mathscr{A}$ (ε, t)-solves the decisional Diffie-Hellman (DDH) problem in the standard model, if $\mathscr{A}$ runs in time t and

$$\left| \Pr[\mathscr{A}(p, g, g^a, g^b, g^{ab}) = 1] - \Pr[\mathscr{A}(p, g, g^a, g^b, g^c) = 1] \right| \geq \varepsilon.$$

We say that an algorithm $\mathscr{A}'$ (ε', t')-solves the DDH problem *semi-generically*, if $\mathscr{A}'$ runs in time t' and

$$\left| \Pr[\mathscr{A}'^{\mathcal{O}(g, g^a, g^b, g^{ab})}(p) = 1] - \Pr[\mathscr{A}'^{\mathcal{O}(g, g^a, g^b, g^c)}(p) = 1] \right| \geq \varepsilon'.$$

It is easy to see that if there exists an algorithm $\mathscr{A}$ (ε, t)-solving DDH in the standard model, then this implies an algorithm $\mathscr{A}'$ (ε', t')-solving DDH semi-generically with success probability $\varepsilon' = \varepsilon$ in time $t' \approx t$:

- $\mathscr{A}'$ makes four reveal-queries to its oracle $\mathscr{O}$ to obtain (g, g^a, g^b, g^d) with $d \in \{ab, c\}$.
- Then $\mathscr{A}'$ runs $\mathscr{A}(p, g, g^a, g^b, g^d)$ as a subroutine.
- $\mathscr{A}'$ outputs whatever $\mathscr{A}$ returns.

Clearly, the success probability of $\mathscr{A}'$ equals the success probability of $\mathscr{A}$, and the running time of $\mathscr{A}'$ equals the running time of $\mathscr{A}$ plus four oracle queries. Thus, an efficient standard-model algorithm implies an efficient semi-generic algorithm for DDH. The converse is trivially true. Thus, when considering the DDH problem, then both the semi-generic group model and the standard model are equivalent.

The above example shows that there exists a common cryptographic problem for which the semi-generic group model is equivalent to the standard model. The next example will show that there exists a strongly related problem, namely the computational version of the decisional problem considered above, for which the semi-generic group model and the standard model seem not to be equivalent.

Example 6.2

Let $\mathbb{G} = (S, \circ)$ be a group of order p with generator g. Let $a, b \xleftarrow{\$} \mathbb{Z}_p$ be uniformly random. We say that an algorithm $\mathscr{A}$ (ε, t)-solves the Computational Diffie-Hellman (CDH) problem in the *standard model*, if $\mathscr{A}$ runs in time t and

$$\Pr[\mathscr{A}(p, g, g^a, g^b) = g^{ab}] \geq \varepsilon.$$

We say that an algorithm $\mathscr{A}'$ (ε', t')-solves the CDH problem *semi-generically*, if $\mathscr{A}'$ runs in time t' and

$$\Pr[\mathscr{A}'^{\mathscr{O}(g, g^a, g^b)}(p) = j : [g^{ab}] = j] \geq \varepsilon'.$$

Thus, in the semi-generic group model we demand that $\mathscr{A}'$ outputs an *index* pointing to a variable containing g^{ab}.

The major difference is that $\mathscr{A}'$ has to compute the solution g^{ab} to its challenge by applying group operations to the initial contents (g, g^a, g^b) of the list L. While all *currently known* algorithms for solving the computational Diffie-Hellman problem are indeed semi-generic, since they solve it by first computing the discrete logarithm of g^a (or g^b), and then computing g^{ab} as $(g^b)^a$ (or $(g^a)^b$), it is not clear that *all* algorithms need to solve CDH this way. Therefore it seems that the semi-generic group model is not equivalent to the standard model when considering CDH. However, note that (in contrast to the black-box generic group model) the *currently known* algorithms, such as index calculus algorithms for finite field groups or elliptic curve groups, are covered in the semi-generic group model.

6.4.2 Analysis of the Diffie-Hellman Problem

As a first application of our model, we show that solving the computational Diffie-Hellman problem semi-generically is as hard as computing discrete logarithms. This simple example illustrates a proving technique which can be applied to other search problems as well.

Theorem 6.1
Let $\mathbb{G} = (S, \circ)$ be a group of prime order p with generator g. Let $\vec{x} = (g, g^a, g^b)$ with $g, g^a, g^b \xleftarrow{\$} S$. Suppose there exists an algorithm $\mathscr{A}$ running in time t and returning an index j such that

$$\Pr[\mathscr{A}^{\mathcal{O}}(p) = j : [g^{ab}] = j] \geq \varepsilon.$$

Then there exists an algorithm $\mathscr{B}$ solving the discrete logarithm in $\mathbb{G}$ in time $t' \approx t$ with success probability $\varepsilon' \geq \varepsilon/2$.

PROOF. Algorithm $\mathscr{B}$ receives as input a discrete logarithm challenge (g, g^y). It implements the semi-generic group oracle $\mathcal{O}$ for $\mathscr{A}$ as follows. $\mathscr{B}$ samples $z \xleftarrow{\$} \mathbb{Z}_p$ and tosses a coin $\rho \xleftarrow{\$} \{0, 1\}$. If $\rho = 0$, then it sets $\vec{x} = (g, g^y, g^z)$. Otherwise $\mathscr{B}$ sets $\vec{x} = (g, g^z, g^y)$. Then it starts $\mathscr{A}$. Clearly, $\mathscr{B}$ can answer all oracle queries of $\mathscr{A}$.

With probability at least ε, $\mathscr{A}$ returns an index j such that $j = [g^{yz}]$. Note that $\mathscr{A}$ needs to compute the list element $L_j = g^{yz}$ by applying a sequence of group operations to the initial contents $\vec{x} = (x_1, x_2, x_3)$ of L, where either $(x_2, x_3) = (g^y, g^z)$ or $(x_2, x_3) = (g^z, g^y)$. Thus, $\mathscr{B}$ obtains an equation

$$g^{yz} = x_1^\alpha \circ x_2^\beta \circ x_3^\gamma = g^\alpha \circ x_2^\beta \circ x_3^\gamma,$$

or equivalently

$$yz \equiv \alpha + \beta \log_g x_2 + \gamma \log_g x_3 \bmod p.$$

Let us consider two cases.

Case 1: $\beta \not\equiv \log_g x_3 \bmod p$. With probability $1/2$ we have $\rho = 0$, and therefore $\log_g x_2 = y$ and $\log_g x_3 = z$. In this case, $\mathscr{B}$ can compute y as

$$yz \equiv \alpha + \beta y + \gamma z \bmod p \iff y \equiv (\alpha + \gamma z)(z - \beta)^{-1} \bmod p.$$

Case 2: $\beta \equiv \log_g x_3 \bmod p$. In this case, $\mathscr{B}$ can trivially compute y, since with probability $1/2$ we have $\rho = 1$, and thus

$$\log_g x_3 \equiv y \equiv \beta \bmod p.$$

Note that in both cases the running time of $\mathscr{B}$ is dominated by the running time of $\mathscr{A}$, plus some minor additional effort to simulate the oracle $\mathscr{O}$. In both cases, $\mathscr{B}$ succeeds with probability at least $\varepsilon/2$. □

The following corollary is essentially an alternative formulation of Theorem 6.1, which captures the most interesting aspect of the result.

Corollary 6.1
Solving the Diffie-Hellman problem semi-generically is equivalent to computing discrete logarithms.

6.5 Semi-Generic Bilinear Groups

Over the last decade a considerable number of innovative cryptosystems, such as identity-based encryption [BF01, BF03], efficient digital signature schemes with strong security [BLS01, BLS04], or powerful primitives like attribute-based encryption [GPSW06, BSW07, OSW07] or functional encryption [OT10, LOS+10] schemes have been proposed over bilinear groups. A bilinear group setting consists of groups $\mathbb{G}_1$, $\mathbb{G}_2$, and $\mathbb{G}_3$, with a bilinear map $e : \mathbb{G}_1 \times \mathbb{G}_2 \rightarrow \mathbb{G}_3$, called a pairing. See Section 2.4 for a description of common types of bilinear group settings with different algebraic properties.

Along with these cryptosystems many new assumptions have been introduced, such as for instance

- bilinear Diffie-Hellman (BDH) [Jou04],

- q-strong Diffie-Hellman [BB04, Che06],

- decision linear Diffie-Hellman (DLIN) [BBS04],

- Co-Diffie-Hellman (Co-DH) [BLS01, BLS04, BGLS03],

and countless more. Unfortunately, for virtually all of them no reduction to a well-analyzed assumption like the discrete logarithm assumption is known. In fact, finding such reductions seems to be a difficult task, since the algebraic settings underlying classic problems (e.g., a single cyclic group for discrete logarithms) significantly differ from bilinear settings. Hence, given an instance of a classic problem it appears to be hard to transform this instance to one of the bilinear problem in order to leverage an algorithm for the former. Consequently, the only way to go beyond pure belief, by providing some sort of immediate evidence that these novel assumptions hold at least heuristically, consists of proofs in restricted models of computation.

So far, the only such model for bilinear settings is a straightforward extension of the generic group model, where all three groups $\mathbb{G}_1$, $\mathbb{G}_2$, and $\mathbb{G}_3$ are modeled as generic groups [RLB$^+$08, KSW08]. We derive two new semi-generic bilinear group models from the general semi-generic model described above, a *weak* semi-generic bilinear group model and a *strong* variant of it.

6.6 Weak Semi-Generic Bilinear Groups

The weak variant is a straightforward extension of the semi-generic group model from Section 6.4 to the bilinear group setting. Let $\mathbb{G}_1 = (S_1, \circ_1)$, $\mathbb{G}_2 = (S_2, \circ_2)$, $\mathbb{G}_3 = (S_3, \circ_3)$ be groups with bilinear pairing $e : \mathbb{G}_1 \times \mathbb{G}_2 \to \mathbb{G}_3$. In the *weak semi-generic bilinear group model* (weak sGGM) an algorithm interacts with an oracle

$$\mathscr{O}(S_1, S_2, S_3, \Pi, \Sigma, \Theta; \vec{x}_1, \vec{x}_2, \vec{x}_3),$$

where $\Pi = \{\circ_1, \circ_2, \circ_3\}$ (in Type 1 or Type 3 settings) or $\Pi = \{\circ_1, \circ_2, \circ_3, \psi\}$ (in Type 2 settings), $\Sigma = \{=\}$, and $\Theta = \{1, 2, 3\}$. Here $=$ denotes the (binary) equality relation.

Note that the algorithm may ask reveal-queries for *all* three groups $\mathbb{G}_1$, $\mathbb{G}_2$, $\mathbb{G}_3$, and thus may obtain all list elements in their standard representation. As we will show below, the techniques used in the previous section to show the semi-generic equivalence of the discrete logarithm and the Diffie-Hellman problem can be adapted to reduce the bilinear discrete logarithm problem to newly introduced computational problems, like Co-DH [BLS01, BLS04, BGLS03].

6.6.1 Relation to the Generic Model and the Standard Model

With the same arguments as in Section 6.4, one can see that the weak semi-generic bilinear group model considers a strictly broader class of algorithms than the generic bilinear group model from Section 2.4, since it allows to exploit specific properties of the given representation of group elements. A well-known example for a computational problem which in the semi-generic bilinear group model is significantly easier than in the generic group model is computing the discrete logarithm problem in bilinear group settings where the MOV-reduction [MOV93] can be applied to solve the problem using index calculus algorithms in $\mathbb{G}_3$.

It is furthermore easy to adopt Examples 6.1 and 6.2 from Section 6.4.1 to bilinear groups, which shows that the weak semi-generic group model is equivalent to the standard model when decisional assumptions are considered, but seems to be stronger when search problems are considered.

6.6.2 Analysis of the Co-Diffie-Hellman Problem

As an application of the weak semi-generic group model, we show that solving the Co-Diffie-Hellman problem semi-generically is equivalent to computing discrete logarithms.

Let $\mathbb{G}_1 = (S_1, \circ_1)$, $\mathbb{G}_2 = (S_2, \circ_2)$, $\mathbb{G}_3 = (S_3, \circ_3)$ be groups of prime order p with bilinear pairing $e : \mathbb{G}_1 \times \mathbb{G}_2 \rightarrow \mathbb{G}_3$. Let us consider an asymmetric bilinear group setting, that is, such that $\mathbb{G}_1 \neq \mathbb{G}_2$. The Co-Diffie-Hellman problem is to compute $g_2^a \in \mathbb{G}_2$, given $g_1, g_1^a \in \mathbb{G}_1$ and $g_2 \in \mathbb{G}_2$. Note that in Type 1 bilinear group settings the Co-Diffie-Hellman is equivalent to the Diffie-Hellman problem. In the semi-generic group model, consider an oracle

$$\mathcal{O}(S_1, S_2, S_3, \Pi, \Sigma, \Theta; \vec{x}_1, \vec{x}_2),$$

where $\vec{x}_1 = (g_1, g_1^a)$ and $\vec{x}_2 = (g_2)$, and either $\Pi = \{\circ_1, \circ_2, \circ_3\}$ (Type 3 setting) or $\Pi = \{\circ_1, \circ_2, \circ_3, \psi\}$ (Type 2 setting).

Definition 6.1
We say that an algorithm $\mathscr{A}$ (ε, t)-solves the Co-Diffie-Hellman problem semi-generically, if $\mathscr{A}$ runs in time t and
$$\Pr[\mathscr{A}^{\mathcal{O}}(p) = j : [g_2^a] = j] \geq \varepsilon.$$

Theorem 6.2
Suppose there exists an algorithm $\mathscr{A}$ that (ε, t)-solves the Co-Diffie-Hellman problem semi-generically. Then there exists an algorithm $\mathscr{B}$ solving the discrete logarithm problem in $\mathbb{G}_1$ in time $t' \approx t$ with success probability $\varepsilon' \geq \varepsilon/2$.

PROOF. Algorithm $\mathscr{B}$ receives as input a discrete logarithm challenge $(g_1, g_1^a) \in S_1 \times S_1$. It implements the semi-generic group oracle $\mathcal{O}$ for $\mathscr{A}$ as follows.

Type 1 setting. This case is identical to the proof of Theorem 6.1 and thus omitted.

Type 2 setting. $\mathscr{B}$ flips a coin $\rho \xleftarrow{\$} \{0, 1\}$. If $\rho = 0$, then $\mathscr{B}$

- computes $h = \psi(g_1)$ and $h^a = \psi(g_1^a)$,

- determines a random generator $g_2 \in S_2$ by sampling $z \xleftarrow{\$} \mathbb{Z}_p^*$ and setting $g_2 := h^z$,

- and sets $\vec{x}_1 = (g_1, g_1^a)$ and $\vec{x}_2 = (g_2)$.

If $\rho = 1$, then $\mathscr{B}$

- samples $z \xleftarrow{\$} \mathbb{Z}_p$ and computes $g_1^z \in \mathbb{G}_1$,

- sets $g_2 := \psi(g_1^a) \in \mathbb{G}_2$,

- and sets $\vec{x}_1 = (g_1, g_1^z)$ and $\vec{x}_2 = (g_2) = (\psi(g_1^a))$.

Then it starts $\mathscr{A}$. Note that in both cases $\mathscr{B}$ can answer all oracle queries of $\mathscr{A}$.

Since $\mathscr{A}$ needs to compute its solution $L_j = g_2^a$ by applying a sequence of group operations to the initial contents of L_2, but in addition may also apply the isomorphism ψ to elements of L_1 and thus can map these elements to L_2, $\mathscr{B}$ obtains an equation

$$L_j = g_2^a = g_2^\alpha h^\beta h^{a\gamma}.$$

where $h = \psi(g_1)$ and the exponents $\alpha, \beta, \gamma \in \mathbb{Z}_p$ can be computed by the oracle by examining the sequence of operations performed by $\mathscr{A}$. Once again we need to consider two cases.

Case 1: $\gamma \not\equiv \log_h g_2 \bmod p$. When $\mathscr{A}$ returns an index j such that $j = [g_2^a]$, then $\mathscr{B}$ obtains an equation

$$L_j = g_2^a = g_2^\alpha h^\beta h^{a\gamma} \iff h^{az} = h^{\alpha z} h^\beta h^{a\gamma},$$

which is equivalent to

$$az \equiv \alpha z + \beta + a\gamma.$$

If $\rho = 0$, then we have $\log_h g_2 = z \bmod p$ and thus $\gamma \not\equiv z \bmod p$. In this case the discrete logarithm a can be computed as

$$a \equiv (\alpha z + \beta)(z - \gamma)^{-1}.$$

Case 2: $\gamma \equiv \log_h g_2 \bmod p$. If $\rho = 1$, then we have $\log_h g_2 = a \bmod p$ and thus $\mathscr{B}$ can compute the discrete logarithm $\gamma \equiv a \bmod p$.

Type 3 setting. $\mathscr{B}$ samples a random generator $g_2 \xleftarrow{\$} S_2$ and sets $\vec{x}_1 = (g_1, g_1^a)$ and $\vec{x}_2 = (g_2)$. Then it runs $\mathscr{A}$. Clearly, $\mathscr{B}$ can answer all oracle queries of $\mathscr{A}$.

With probability at least ε, $\mathscr{A}$ returns an index j such that $j = [g_2^a]$. Algorithm $\mathscr{A}$ needs to compute the list element $L_j = g_2^a$ by applying a sequence of group operations to the initial contents $\vec{x}_2 = (g_2)$ of L_2. Thus, $\mathscr{B}$ can observe the sequence of group operations performed by $\mathscr{A}$ to obtain $\alpha \in \mathbb{Z}_p$ such that $g_2^a = g_2^\alpha$, which is equivalent to $\alpha \equiv a \bmod p$.

$\square$

6.7 Strong Semi-Generic Bilinear Groups

As noted before, the weak semi-generic group model is equivalent to the standard model (and thus not useful) when decisional assumptions are considered. Therefore we introduce a second, stronger model. In this model an algorithm interacts with an oracle

$$\mathscr{O}((S_1, S_2, S_3), \Pi, \Sigma, \Theta; \vec{x}_1, \vec{x}_2, \vec{x}_3),$$

where $\Sigma = \{=_1, =_2, =_3\}$, $\Pi = \{\circ_1, \circ_2, \circ_3, e\}$, and $\Theta = \{3\}$. Note that in contrast to the weak model, we have $\Theta = \{3\}$ instead of $\Theta = \{1, 2, 3\}$. Thus, the algorithm may only obtain elements of $\mathbb{G}_3$ in their standard representation, while $\mathbb{G}_1$ and $\mathbb{G}_2$ are modeled as generic groups. The justification for this model is based on the following two observations.

1. In *all known instances* of bilinear settings, the groups $\mathbb{G}_1$ and $\mathbb{G}_2$ are elliptic curve groups. Modeling elliptic curve groups as generic groups may be considered as a reasonable abstraction.

2. However, in contrast to that, the group $\mathbb{G}_3$ is usually a subgroup of the multiplicative group of a finite field. Modeling this group as a generic group is clearly inappropriate: There exist algorithms, which exploit the field structure of $\mathbb{G}_3$ to solve cryptographic problems siginicantly faster than generic group algorithms. For instance, there are sub-exponential time algorithms that map the inputs over $\mathbb{G}_1$ and $\mathbb{G}_2$ (given as part of a problem instance) to $\mathbb{G}_3$, using the bilinear map, and then determine the discrete logarithms of these elements over $\mathbb{G}_3$ using index calculus (MOV reduction [MOV93]).

The stronger model leverages this observation by modeling the elliptic curve groups as black-box generic groups, while reveal-queries are allowed for $\mathbb{G}_3$. Thus, an algorithm can perform *any* computation over $\mathbb{G}_3$ that is possible in the subgroup of a finite field. This enables us to relate the semi-generic hardness of many decisional problems from pairing based cryptography, like the bilinear decisional Diffie-Hellman problem (BDDH), to one single hardness assumption in $\mathbb{G}_3$, namely the *square decisional Diffie-Hellman* assumptions.

6.7.1 Relation to the Generic Group Model and to the Weak Semi-Generic Group Model

Again it is easy to see that the strong semi-generic bilinear group model considers a strictly broader class of algorithms than the generic bilinear group model from Section 2.4, since at least specific properties of the given representation of elements of $\mathbb{G}_3$ can be exploited. This is still sufficient to capture known attacks, such as the MOV-reduction [MOV93], which are excluded in the generic group model.

For instance, the techniques of Dent [Den02] can be used to separate the strong model from the weak model. The (contrived but valid) separation between generic groups and standard groups from [Den02] adopts the separation of Canetti, Goldreich, and Halevi [CGH98, CGH04] from random oracles to generic groups.

The strong model is clearly more restrictive than its weak sibling, thus a smaller class of algorithms is considered, but both the weak and the strong model are significantly closer to the "real world" than the generic group model. In general, it is a more realistic heuristic to assume that a given bilinear group setting realizes weak rather than strong semi-generic groups, therefore the weak model should be preferred over the strong variant. However, to the best of our knowledge *all algorithms currently known for solving pairing-based algebraic problems work also in the strong semi-generic group model*. This holds in particular for the subexponential time algorithms applying a MOV reduction [MOV93]. Moreover, an important justification for the strong model is that it allows to analyze decisional assumptions, as we illustrate below.

6.7.2 Some Helpful Observations

This section describes a few simple observations that will turn out to be the helpful ingredients for proofs in the strong semi-generic group model.

Observation 1: The internal representation of elements inside the oracle is arbitrary. In the strong semi-generic group model, an algorithm may obtain only elements of $\mathbb{G}_3$ in their explicit representation. Elements of $\mathbb{G}_1$ and $\mathbb{G}_2$ are only represented implicitly by pointers referring to list entries. Thus, algorithms are "blind" with respect to the internal representation of the groups $\mathbb{G}_1$ and $\mathbb{G}_2$ (as well as the pairing e and the isomorphism ψ).

Hence, we can plug in "something else" for these components as long as these replacements behave like cyclic groups with a bilinear map and an isomorphism. We will utilize this observation to map inputs given over $\mathbb{G}_3$ back to $\mathbb{G}_1$ and $\mathbb{G}_2$ by setting $\mathbb{G}_1 := \mathbb{G}_2 := \mathbb{G}_3$ internally, and simulating the bilinear map e and the isomorphism ψ.

Observation 2: Computed elements over $\mathbb{G}_1$ and $\mathbb{G}_2$ are linear polynomials in initial inputs. Consider a strong semi-generic group oracle $\mathscr{O}(\vec{x}_1, \vec{x}_2, \vec{x}_3)$, where $\vec{x}_1 = (x_{1,1}, \ldots, x_{1,m})$ and $\vec{x}_2 = (x_{2,1}, \ldots, x_{2,n})$. A semi-generic algorithm $\mathscr{A}$ is restricted to applying a sequence of group operations to elements of the list L_1 and L_2. Thus, in Type 1 and Type 3 settings (without efficiently computable isomorphism $\psi : \mathbb{G}_1 \to \mathbb{G}_2$), each element of L_1 and L_2 computed by the algorithm can be written as

$$L_{1,j} = \prod_{k=1}^{m} x_{1,k}^{\alpha_{1,k}} \quad \text{and} \quad L_{2,j} = \prod_{k=1}^{n} x_{2,k}^{\alpha_{2,k}}$$

for integers $\alpha_{i,k}$ that can be computed by the oracle by examining the sequence of operations performed by $\mathscr{A}$. In Type 2 settings we additionally have the isomorphism $\psi : \mathbb{G}_1 \to \mathbb{G}_2$. In this case, each list element computed by the algorithm can be written as

$$L_{1,j} = \prod_{k=1}^{m} x_{1,k}^{\alpha_{1,k}} \quad \text{and} \quad L_{2,j} = \prod_{k=1}^{n} x_{2,k}^{\alpha_{2,k}} \prod_{k=1}^{m} \psi(x_{1,k})^{\beta_{2,k}}$$

for known integers $\alpha_{i,k}$ and $\beta_{2,k}$.

Note that Observation 2 has already been used in the proofs of Theorems 6.1 and 6.2. However, we think it is instructive to state it explicitly here, since it simplifies the description of Observation 3.

Observation 3: The pairing is simulatable knowing the images of initial inputs. Let $L_{1,i} \in \mathbb{G}_1$ and $L_{2,j} \in \mathbb{G}_2$ be two elements computed by a semi-generic algorithm. Then by using Observation 2 it is easy to see that in Type 1 and Type 3 settings we have

$$e(L_{1,i}, L_{2,j}) = e(\prod_{k=1}^{m} x_{1,k}^{\alpha_{1,k}}, \prod_{\ell=1}^{n} x_{2,\ell}^{\alpha_{2,\ell}}) = \prod_{k=1}^{m} \prod_{\ell=1}^{n} e(x_{1,k}, x_{2,\ell})^{\alpha_{1,\ell} \alpha_{2,\ell}},$$

where the integers $\alpha_{i,j}$ can be computed by examining the sequence of operations performed by the algorithm. Thus, in order to be able to simulate the pairing map for all elements computed by the algorithm, it suffices to know all elements of the set

$$\Theta = \left\{ e(x_{1,k}, x_{2,\ell}) : k \in [m], \ell \in [n] \right\}$$

containing the images of all initial inputs to the oracle under the pairing map. The case of Type 2 settings is analogous, howevever, we need to take the isomorphism map into account. Thus we need

$$\Theta' = \Theta \cup \left\{ e(x_{1,k}, \psi(x_{1,\ell})) : k, \ell \in [m] \right\}.$$

6.7.3 Analysis of the Decisional Bilinear Diffie-Hellman Problem

The bilinear decisional Diffie-Hellman problem (bilinear DDH) is certainly one of the most important computational problems over bilinear groups. It has originally been introduced in a seminal paper by Joux [Jou00, Jou04] and since then been used in several cryptographic constructions, such as [BF01, BF03, Wat05, KG06, BDNS07] and many more.

Definition. Let us consider the bilinear DDH problem over a Type 1 setting. Consider groups $\mathbb{G}_1 = (S_1, \circ_1)$ and $\mathbb{G}_3 = (S_3, \circ_3)$ of prime order p with generator $g_1 \in \mathbb{G}_1$ and bilinear map $e : \mathbb{G}_1 \times \mathbb{G}_1 \to \mathbb{G}_3$. Let $g_3 = e(g_1, g_1)$. The bilinear DDH problem is to compute $b \in \{0,1\}$ on input $(g_1, g_1^x, g_1^y, g_1^z, g_3^{r_b})$, where $x, y, z, r_0 \xleftarrow{\$} \mathbb{Z}_p$, $r_1 = xyz$, and $b \xleftarrow{\$} \{0,1\}$. In the strong semi-generic group model, this problem can be defined as follows. Consider a strong semi-generic bilinear group oracle

$$\mathscr{O}(\vec{x}_1, \vec{x}_3),$$

where $\vec{x}_1 = (g_1, g_1^x, g_1^y, g_1^z)$ and $\vec{x}_3 = (g_3^{r_b})$.

Definition 6.2
We say the that the bilinear decisional Diffie-Hellman problem is $(\varepsilon_{\text{bDDH}}, t)$-hard in the strong semi-generic model, if

$$\left| \Pr[\mathscr{A}^{\mathscr{O}((g_1, g_1^x, g_1^y, g_1^z),(g_3^{xyz}))}(p) = 1] - \Pr[\mathscr{A}^{\mathscr{O}((g_1, g_1^x, g_1^y, g_1^z),(g_3^{r_0}))}(p) = 1] \right| \leq \varepsilon_{\text{bDDH}}.$$

for all algorithms $\mathscr{A}$ running in time t.

Square Decisional Diffie-Hellman. We relate the semi-generic hardness of the bilinear DDH problem to the hardness of the square decisional Diffie-Hellman (square DDH) problem over $\mathbb{G}_3$. The square DDH problem is a potentially easier variant of the decisional Diffie-Hellman problem: Given $(g_3, g_3^x, g_3^{r_b})$, where $x, r_0 \xleftarrow{\$} \mathbb{Z}_p$, $b \xleftarrow{\$} \{0, 1\}$, and $r_1 = x^2$, compute b.

Definition 6.3
We say that the square decisional Diffie-Hellman problem is $(\varepsilon_{\text{sqDDH}}, t)$-hard, if

$$\left| \Pr[\mathscr{A}(p, g_3, g_3^x, g_3^{x^2}) = 1] - \Pr[\mathscr{A}(p, g_3, g_3^x, g_3^{r_0}) = 1] \right| \leq \varepsilon_{\text{sqDDH}}$$

for all algorithms $\mathscr{A}$ running in time t.

An auxiliary lemma. As a last prerequisite we need the following lemma.

Lemma 6.1
Let $\mathbb{G}_3$ be a group of prime order p with random generator g_3. Suppose that solving the square DDH problem in $\mathbb{G}_3$ is $(\varepsilon_{\text{sqDDH}}, t)$-hard. Let

$$V_0 := (g_3, g_3^x, g_3^y, g_3^z, g_3^{x^2}, g_3^{y^2}, g_3^{z^2}, g_3^{xy}, g_3^{xz}, g_3^{yz}, g_3^r),$$
$$V_1 := (g_3, g_3^x, g_3^y, g_3^z, g_3^{x^2}, g_3^{y^2}, g_3^{z^2}, g_3^{xy}, g_3^{xz}, g_3^{yz}, g_3^{xyz})$$

for random $x, y, z, r \xleftarrow{\$} \mathbb{Z}_p$. Then for all algorithms $\mathscr{A}$ running in time $t' \approx t$ holds that

$$|\Pr[\mathscr{A}(p, V_0) = 1] - \Pr[\mathscr{A}(p, V_1) = 1]| \leq 9 \cdot \varepsilon_{\text{sqDDH}}.$$

The proof of Lemma 6.1 is deferred to Section 6.7.4.

Main Theorem. Now we are ready to state and prove the main theorem of this section.

Theorem 6.3
If solving the square DDH problem over $\mathbb{G}_3$ is $(\varepsilon_{\text{sqDDH}}, t)$-hard, then solving the bilinear DDH problem semi-generically is $(\varepsilon_{\text{bDDH}}, t')$-hard with

$$\varepsilon_{\text{bDDH}} \leq 9 \cdot \varepsilon_{\text{sqDDH}} \quad \text{and} \quad t \approx t'.$$

PROOF. We proceed in a sequence of four games $\mathbb{G}_1, \ldots, \mathbb{G}_3$. In the first game, we consider an interaction of an adversary $\mathscr{A}$ with an oracle with proceeds exactly like the oracle $\mathscr{O}((g_1, g_1^x, g_1^y, g_1^z), (g_3^{xyz}))$ described above, when it receives as input a "real" bilinear DDH tuple. Then we gradually make modifications to this oracle, until we end up with the last game where we have an oracle proceeding exactly like oracle $\mathcal{O}((g_1, g_1^x, g_1^y, g_1^z), (g_3^r))$ receiving as input a "random" bilinear DDH tuple with $\vec{x}_1 = (g_1, g_1^x, g_1^y, g_1^z)$ and $\vec{x}_3 = (g_3^r)$ for random $x, y, z, r \xleftarrow{\$} \mathbb{Z}_p$ as input.

In the following let $\mathcal{O}_i$ denote the oracle that $\mathscr{A}$ interacts with in Game i. Thus, to prove our claim we have to show that

$$\left| \Pr[\mathscr{A}^{\mathcal{O}_0} = 1] - \Pr[\mathscr{A}^{\mathcal{O}_3} = 1] \right| \leq 9 \cdot \varepsilon_{\mathsf{sqDDH}}.$$

Game 0.   In this game, the oracle $\mathcal{O}_0$ proceeds exactly like the original oracle

$$\mathcal{O}((g_1, g_1^x, g_1^y, g_1^z), (g_3^{xyz})).$$

Game 1.   In this game we modify the internal representation of elements of $\mathbb{G}_1$ inside the oracle. Technically, we replace $\mathcal{O}_0$ with an oracle $\mathcal{O}_1$, which receives as input a single vector

$$\vec{x} = (g_3, g_3^x, g_3^y, g_3^z, g_3^{x^2}, g_3^{y^2}, g_3^{z^2}, g_3^{xy}, g_3^{xz}, g_3^{yz}, g_3^{xyz})$$

containing only elements of $\mathbb{G}_3$, where g_3 is a random generator of $\mathbb{G}_3$ and $x, y, z \xleftarrow{\$} \mathbb{Z}_p$ are random integers. $\mathcal{O}_1$ simulates $\mathcal{O}_0$ as follows.

- It initializes the list L_1 with $(g_3, g_3^x, g_3^y, g_3^z)$, and list L_3 with (g_3^{xyz}).

- When queried to apply a group operation or equality test of elements of L_1 or L_3, it applies the queried operation to elements of $\mathbb{G}_3$. Note that $\mathbb{G}_1$ and $\mathbb{G}_3$ are cyclic groups of equal order (and thus isomorphic), and that the internal representation of elements of $\mathbb{G}_1$ inside the black-box is hidden from the algorithm (cf. Observation 1 in Section 6.7.2).

- When queried to apply the bilinear map to two elements of L_1, it uses the additional group elements $g^{x^2}, g^{y^2}, g^{z^2}, g_3^{xy}, g_3^{xz}, g_3^{yz}$ from its input to run the simulation procedure illustrated in Observation 3 from Section 6.7.2. The result is appended to L_3.

- Reveal-queries to elements of L_3 are answered as before.

The change of the internal representation of group elements is hidden from the adversary, since the group operations and the pairing evaluation can be simulated. Thus, for algorithm $\mathscr{A}$ the oracle $\mathcal{O}_1$ is perfectly indistinguishable from $\mathcal{O}_0$, which yields

$$\Pr[\mathscr{A}^{\mathcal{O}_1} = 1] = \Pr[\mathscr{A}^{\mathcal{O}_0} = 1].$$

Game 2. We modify the input to $\mathcal{O}_1$. Oracle $\mathcal{O}_2$ proceeds exactly like oracle prevgame, except that it receives as input a vector

$$\vec{x} = (g_3, g_3^x, g_3^y, g_3^z, g_3^{x^2}, g_3^{y^2}, g_3^{z^2}, g_3^{xy}, g_3^{xz}, g_3^{yz}, g_3^r)$$

where g_3 is a random generator of $\mathbb{G}_3$ and $x, y, z, r \xleftarrow{\$} \mathbb{Z}_p$ are random integers. Note that the input vector is identical to the input vector from Game 1, except that the last element is g_3^r for random r instead of g_3^{xyz}. By Lemma 6.1, we have

$$\left| \Pr[\mathscr{A}^{\mathcal{O}_2} = 1] - \Pr[\mathscr{A}^{\mathcal{O}_1} = 1] \right| \leq 9 \cdot \varepsilon_{\mathsf{sqDDH}}.$$

Game 3. In this game, we revert our modification introduced in the transition from Game 0 to Game 1. That is, our Oracle $\mathcal{O}_3$ proceeds exactly like oracle $\mathcal{O}((g_1, g_1^x, g_1^y, g_1^z), (g_3^r))$.

With the same arguments as in Game 1, one can see that oracle $\mathcal{O}_2$ from the Game 2 is a perfect simulator for Oracle $\mathcal{O}((g_1, g_1^x, g_1^y, g_1^z), (g_3^r))$, and thus a perfect simulator for $\mathcal{O}_3$, whose only essential difference is that it uses a different internal representation of group elements. This implies

$$\Pr[\mathscr{A}^{\mathcal{O}_3} = 1] = \Pr[\mathscr{A}^{\mathcal{O}_2} = 1].$$

Summing up probabilities from Game 0 to Game 3, we obtain

$$\left| \Pr[\mathscr{A}^{\mathcal{O}_0} = 1] - \Pr[\mathscr{A}^{\mathcal{O}_3} = 1] \right| \leq 9 \cdot \varepsilon_{\mathsf{sqDDH}}.$$

$\square$

6.7.4 Proof of Lemma 6.1

To complete the proof of our theorem on the semi-generic hardness of bilinear decisional Diffie-Hellman, it remains to prove Lemma 6.1. To this end, we first need to define the decisional Diffie-Hellman problem and show that the square decisional Diffie-Hellman problem is a potentially easier variant of it.

Decisional Diffie-Hellman and its Relation to Square DDH. The decisional Diffie-Hellman (DDH) problem is to compute $b \in \{0, 1\}$, given

$$(g_3, g_3^x, g_3^y, g_3^{r_b}),$$

where $a, b, r_0 \xleftarrow{\$} \mathbb{Z}_p$, $r_1 = xy$, and $b \xleftarrow{\$} \{0, 1\}$.

Definition 6.4

Let $\mathbb{G}_3$ be a group with random generator g_3 of order p. Let $x, y, r_0 \xleftarrow{\$} \mathbb{Z}_p$ be random integers. We say that the decisional Diffie-Hellman problem is $(\varepsilon_{\mathsf{DDH}}, t)$-hard, if

$$\left| \Pr[\mathscr{A}(p, g_3, g_3^x, g_3^y, g_3^{xy}) = 1] - \Pr[\mathscr{A}(p, g_3, g_3^x, g_3^y, g_3^{r_0}) = 1] \right| \leq \varepsilon_{\mathsf{DDH}}$$

for all algorithms $\mathscr{A}$ running in time t.

It is a simple fact that the square DDH problem reduces to the DDH problem, thus square DDH is potentially easier than DDH. This is captured by the following lemma. It is unknown whether both problems are equivalent.

Lemma 6.2

If solving the square DDH problem is $(\varepsilon_{\mathsf{sqDDH}}, t)$-hard, then solving the DDH problem is $(\varepsilon_{\mathsf{DDH}}, t')$-hard with $t \approx t'$ and $\varepsilon_{\mathsf{DDH}} \leq \varepsilon_{\mathsf{sqDDH}}$.

PROOF. We suppose there exists an algorithm $\mathscr{A}$ $(\varepsilon_{\mathsf{DDH}}, t')$-solving the DDH problem, and show that this implies an algorithm $\mathscr{B}$ that (ε, t')-solves the square DDH problem with $t' \approx t$.

$\mathscr{B}$ receives as input a square DDH challenge $(g_3, g_3^x, g_3^{r_b})$. It samples a random integer $\alpha \xleftarrow{\$} \mathbb{Z}_p$, sets $h := g_3^{x+\alpha}$ and $h' := g_3^{r_b + x\alpha}$, and runs $\mathscr{A}(p, g_3, g_3^x, h, h')$. Algorithm $\mathscr{B}$ returns whatever $\mathscr{A}$ returns.

Note that if $g_3^{r_b} = g_3^{x^2}$, then we have $h' = g_3^{x^2 + x\alpha} = g_3^{x(x+\alpha)}$. In this case, $\mathscr{A}$ receives a correctly distributed "real" DDH tuple as input. If $g_3^{r_b}$ is a uniformly random group element, then $h' := g_3^{r_b + x\alpha}$ is uniformly random, thus $\mathscr{A}$ receives a correctly distributed "random" DDH tuple. Thus $\mathscr{B}$ can use $\mathscr{A}$ to solve its square DDH challenge. $\qquad\square$

Proof of Lemma 6.1. To prove Lemma 6.1, consider the following sequence of vectors of group elements, where each vector W_i is identical to W_{i-1} except for the entries highlighted in boldface.

$$W_0 := (g_3, g_3^x, g_3^y, g_3^z, g_3^{x^2}, g_3^{y^2}, g_3^{z^2}, g_3^{xy}, g_3^{xz}, g_3^{yz}, g_3^r),$$

$$W_1 := (g_3, g_3^x, g_3^y, g_3^z, \mathbf{g_3^{r_1}}, g_3^{y^2}, g_3^{z^2}, g_3^{xy}, g_3^{xz}, g_3^{yz}, g_3^r),$$

$$W_2 := (g_3, g_3^x, g_3^y, g_3^z, g_3^{r_1}, \mathbf{g_3^{r_2}}, g_3^{z^2}, g_3^{xy}, g_3^{xz}, g_3^{yz}, g_3^r),$$

$$W_3 := (g_3, g_3^x, g_3^y, g_3^z, g_3^{r_1}, g_3^{r_2}, \mathbf{g_3^{r_3}}, g_3^{xy}, g_3^{xz}, g_3^{yz}, g_3^r),$$

$$W_4 := (g_3, g_3^x, g_3^y, g_3^z, g_3^{r_1}, g_3^{r_2}, g_3^{r_3}, \mathbf{g_3^{r_4}}, g_3^{xz}, g_3^{yz}, g_3^r),$$

$$W_5 := (g_3, g_3^x, g_3^y, g_3^z, g_3^{r_1}, g_3^{r_2}, g_3^{r_3}, g_3^{r_4}, g_3^{xz}, g_3^{yz}, \mathbf{g_3^{r_4 z}}),$$

$$W_6 := (g_3, g_3^x, g_3^y, g_3^z, g_3^{r_1}, g_3^{r_2}, g_3^{r_3}, \mathbf{g_3^{xy}}, g_3^{xz}, g_3^{yz}, \mathbf{g_3^{xyz}}),$$

$$W_7 := (g_3, g_3^x, g_3^y, g_3^z, \mathbf{g_3^{x^2}}, g_3^{r_2}, g_3^{r_3}, g_3^{xy}, g_3^{xz}, g_3^{yz}, g_3^{xyz}),$$

$$W_8 := (g_3, g_3^x, g_3^y, g_3^z, g_3^{x^2}, \mathbf{g_3^{y^2}}, g_3^{r_3}, g_3^{xy}, g_3^{xz}, g_3^{yz}, g_3^{xyz}),$$

$$W_9 := (g_3, g_3^x, g_3^y, g_3^z, g_3^{x^2}, g_3^{y^2}, \mathbf{g_3^{z^2}}, g_3^{xy}, g_3^{xz}, g_3^{yz}, g_3^{xyz}).$$

First, note that W_0 is identically distributed to V_0, and W_9 is distributed exactly like V_1. Moreover, observe that for all $i \in \{1,2,3,7,8,9\}$ an algorithm distinguishing W_i from W_{i-1} in time t with advantage ε implies an algorithm solving the square DDH problem in time $t' \approx t$ with advantage at least $\varepsilon_{\mathsf{sqDDH}} \geq \varepsilon$. Likewise, for all $i \in \{4,5,6\}$ an algorithm distinguishing W_i from W_{i-1} in time t with advantage ε implies an algorithm solving the DDH problem in time $t' \approx t$ with advantage at least $\varepsilon_{\mathsf{DDH}} \geq \varepsilon$. Thus, in combination with Lemma 6.2 we have

$$\varepsilon \leq 6 \cdot \varepsilon_{\mathsf{sqDDH}} + 3 \cdot \varepsilon_{\mathsf{DDH}} \implies \varepsilon \leq 9 \cdot \varepsilon_{\mathsf{sqDDH}}.$$

Bibliography

[AJR08]   ALTMANN, KRISTINA, TIBOR JAGER and ANDY RUPP: *On Black-Box Ring Extraction and Integer Factorization*. In ACETO, LUCA, IVAN DAMGÅRD, LESLIE ANN GOLDBERG, MAGNÚS M. HALLDÓRSSON, ANNA INGÓLFSDÓTTIR and IGOR WALUKIEWICZ (editors): *ICALP 2008: 35th International Colloquium on Automata, Languages and Programming, Part II*, volume 5126 of *Lecture Notes in Computer Science*, pages 437–448. Springer, July 2008.

[AM09]   AGGARWAL, DIVESH and UELI MAURER: *Breaking RSA Generically Is Equivalent to Factoring*. In JOUX, ANTOINE (editor): *Advances in Cryptology – EUROCRYPT 2009*, volume 5479 of *Lecture Notes in Computer Science*, pages 36–53. Springer, April 2009.

[AMS11]   AGGARWAL, DIVESH, UELI MAURER and IGOR SHPARLINSKI: *The Equivalence of Strong RSA and Factoring in the Generic Ring Model of Computation*. In AUGOT, DANIEL and ANNE CANTEAUT (editors): *Workshop on Coding and Cryptography - WCC 2011*. INRIA, July 2011.

[BB04]   BONEH, DAN and XAVIER BOYEN: *Short Signatures Without Random Oracles*. In CACHIN, CHRISTIAN and JAN CAMENISCH (editors): *Advances in Cryptology – EUROCRYPT 2004*, volume 3027 of *Lecture Notes in Computer Science*, pages 56–73. Springer, May 2004.

[BB08]   BONEH, DAN and XAVIER BOYEN: *Short Signatures Without Random Oracles and the SDH Assumption in Bilinear Groups*. Journal of Cryptology, 21(2):149–177, April 2008.

[BBS04]   BONEH, DAN, XAVIER BOYEN and HOVAV SHACHAM: *Short Group Signatures*. In FRANKLIN, MATTHEW (editor): *Advances in Cryptology – CRYPTO 2004*, volume 3152 of *Lecture Notes in Computer Science*, pages 41–55. Springer, August 2004.

[BDNS07]   BIRKETT, JAMES, ALEXANDER W. DENT, GREGORY NEVEN
           and JACOB C. N. SCHULDT: *Efficient Chosen-Ciphertext Secure
           Identity-Based Encryption with Wildcards*. In PIEPRZYK, JOSEF,
           HOSSEIN GHODOSI and ED DAWSON (editors): *ACISP 07: 12th
           Australasian Conference on Information Security and Privacy*, vol-
           ume 4586 of *Lecture Notes in Computer Science*, pages 274–292.
           Springer, July 2007.

[BF01]   BONEH, DAN and MATTHEW K. FRANKLIN: *Identity-Based En-
           cryption from the Weil Pairing*. In KILIAN, JOE (editor): *Advances
           in Cryptology – CRYPTO 2001*, volume 2139 of *Lecture Notes in
           Computer Science*, pages 213–229. Springer, August 2001.

[BF03]   BONEH, DAN and MATTHEW K. FRANKLIN: *Identity Based En-
           cryption from the Weil Pairing*. SIAM Journal on Computing,
           32(3):586–615, 2003.

[BGLS03]   BONEH, DAN, CRAIG GENTRY, BEN LYNN and HOVAV
           SHACHAM: *Aggregate and Verifiably Encrypted Signatures from
           Bilinear Maps*. In BIHAM, ELI (editor): *Advances in Cryptology
           – EUROCRYPT 2003*, volume 2656 of *Lecture Notes in Computer
           Science*, pages 416–432. Springer, May 2003.

[BL96]   BONEH, DAN and RICHARD J. LIPTON: *Algorithms for Black-
           Box Fields and their Application to Cryptography (Extended Ab-
           stract)*. In KOBLITZ, NEAL (editor): *Advances in Cryptology –
           CRYPTO'96*, volume 1109 of *Lecture Notes in Computer Science*,
           pages 283–297. Springer, August 1996.

[Bla06]   BLACK, JOHN: *The Ideal-Cipher Model, Revisited: An Uninstan-
           tiable Blockcipher-Based Hash Function*. In ROBSHAW, MATTHEW
           J. B. (editor): *Fast Software Encryption – FSE 2006*, volume 4047
           of *Lecture Notes in Computer Science*, pages 328–340. Springer,
           March 2006.

[BLS01]   BONEH, DAN, BEN LYNN and HOVAV SHACHAM: *Short Signa-
           tures from the Weil Pairing*. In BOYD, COLIN (editor): *Advances in
           Cryptology – ASIACRYPT 2001*, volume 2248 of *Lecture Notes in
           Computer Science*, pages 514–532. Springer, December 2001.

[BLS04]   BONEH, DAN, BEN LYNN and HOVAV SHACHAM: *Short Signatures from the Weil Pairing*. Journal of Cryptology, 17(4):297–319, September 2004.

[Boy08]   BOYEN, X.: *The Uber-Assumption Family*. In GALBRAITH, STEVEN D. and KENNETH G. PATERSON (editors): *Pairing*, volume 5209 of *Lecture Notes in Computer Science*, pages 39–56. Springer, 2008.

[BR93]   BELLARE, MIHIR and PHILLIP ROGAWAY: *Random Oracles are Practical: A Paradigm for Designing Efficient Protocols*. In ASHBY, V. (editor): *ACM CCS 93: 1st Conference on Computer and Communications Security*, pages 62–73. ACM Press, November 1993.

[BR06]   BELLARE, MIHIR and PHILLIP ROGAWAY: *The Security of Triple Encryption and a Framework for Code-Based Game-Playing Proofs*. In VAUDENAY, SERGE (editor): *Advances in Cryptology – EUROCRYPT 2006*, volume 4004 of *Lecture Notes in Computer Science*, pages 409–426. Springer, May / June 2006.

[Bro05]   BROWN, DANIEL R. L.: *Breaking RSA May Be As Difficult As Factoring*. Cryptology ePrint Archive, Report 2005/380, 2005. http://eprint.iacr.org/.

[BS84]   BABAI, LÁSZLÓ and ENDRE SZEMERÉDI: *On the Complexity of Matrix Group Problems I*. In *FOCS*, pages 229–240. IEEE, 1984.

[BSW07]   BETHENCOURT, JOHN, AMIT SAHAI and BRENT WATERS: *Ciphertext-Policy Attribute-Based Encryption*. In *2007 IEEE Symposium on Security and Privacy*, pages 321–334. IEEE Computer Society Press, May 2007.

[BV98]   BONEH, DAN and RAMARATHNAM VENKATESAN: *Breaking RSA May Not Be Equivalent to Factoring*. In NYBERG, KAISA (editor): *Advances in Cryptology – EUROCRYPT'98*, volume 1403 of *Lecture Notes in Computer Science*, pages 59–71. Springer, May / June 1998.

[CFW11]   CATALANO, DARIO, DARIO FIORE and BOGDAN WARINSCHI: *Adaptive Pseudo-free Groups and Applications*. In PATERSON, KENNETH G. (editor): *Advances in Cryptology – EURO-*

CRYPT 2011, volume 6632 of *Lecture Notes in Computer Science*, pages 207–223. Springer, May 2011.

[CGH98]   CANETTI, RAN, ODED GOLDREICH and SHAI HALEVI: *The Random Oracle Methodology, Revisited (Preliminary Version)*. In *30th Annual ACM Symposium on Theory of Computing*, pages 209–218. ACM Press, May 1998.

[CGH04]   CANETTI, RAN, ODED GOLDREICH and SHAI HALEVI: *The random oracle methodology, revisited*. J. ACM, 51(4):557–594, 2004.

[CGHGN01] CATALANO, DARIO, ROSARIO GENNARO, NICK HOWGRAVE-GRAHAM and PHONG Q. NGUYEN: *Paillier's Cryptosystem Revisited*. In *ACM CCS 01: 8th Conference on Computer and Communications Security*, pages 206–214. ACM Press, November 2001.

[Che06]   CHEON, JUNG HEE: *Security Analysis of the Strong Diffie-Hellman Problem*. In VAUDENAY, SERGE (editor): *Advances in Cryptology – EUROCRYPT 2006*, volume 4004 of *Lecture Notes in Computer Science*, pages 1–11. Springer, May / June 2006.

[CNS02]   CATALANO, DARIO, PHONG Q. NGUYEN and JACQUES STERN: *The Hardness of Hensel Lifting: The Case of RSA and Discrete Logarithm*. In ZHENG, YULIANG (editor): *Advances in Cryptology – ASIACRYPT 2002*, volume 2501 of *Lecture Notes in Computer Science*, pages 299–310. Springer, December 2002.

[DBS04]   DUTTA, RATNA, RANA BARUA and PALASH SARKAR: *Pairing-Based Cryptographic Protocols : A Survey*. Cryptology ePrint Archive, Report 2004/064, 2004. http://eprint.iacr.org/.

[den90]   DEN BOER, BERT: *Diffie-Hellman is as Strong as Discrete Log for Certain Primes (Rump Session)*. In GOLDWASSER, SHAFI (editor): *Advances in Cryptology – CRYPTO '88*, volume 403 of *Lecture Notes in Computer Science*, pages 530–539. Springer, August 1990.

[Den02]   DENT, ALEXANDER W.: *Adapting the Weaknesses of the Random Oracle Model to the Generic Group Model*. In ZHENG, YULIANG (editor): *Advances in Cryptology – ASIACRYPT 2002*, volume 2501 of *Lecture Notes in Computer Science*, pages 100–109. Springer, December 2002.

[DH76]   DIFFIE, WHITFIELD and MARTIN HELLMAN: *New directions in cryptography*. IEEE Transactions on Information Theory, 22:644–654, 1976.

[DJ01]   DAMGÅRD, IVAN and MATS JURIK: *A Generalisation, a Simplification and Some Applications of Paillier's Probabilistic Public-Key System*. In KIM, KWANGJO (editor): *PKC 2001: 4th International Workshop on Theory and Practice in Public Key Cryptography*, volume 1992 of *Lecture Notes in Computer Science*, pages 119–136. Springer, February 2001.

[DK02]   DAMGÅRD, IVAN and MACIEJ KOPROWSKI: *Generic Lower Bounds for Root Extraction and Signature Schemes in General Groups*. In KNUDSEN, LARS R. (editor): *Advances in Cryptology – EUROCRYPT 2002*, volume 2332 of *Lecture Notes in Computer Science*, pages 256–271. Springer, April / May 2002.

[FGK+10]   FREEMAN, DAVID MANDELL, ODED GOLDREICH, EIKE KILTZ, ALON ROSEN and GIL SEGEV: *More Constructions of Lossy and Correlation-Secure Trapdoor Functions*. In NGUYEN, PHONG Q. and DAVID POINTCHEVAL (editors): *PKC 2010: 13th International Conference on Theory and Practice of Public Key Cryptography*, volume 6056 of *Lecture Notes in Computer Science*, pages 279–295. Springer, May 2010.

[Fis00]   FISCHLIN, MARC: *A Note on Security Proofs in the Generic Model*. In OKAMOTO, TATSUAKI (editor): *Advances in Cryptology – ASIACRYPT 2000*, volume 1976 of *Lecture Notes in Computer Science*, pages 458–469. Springer, December 2000.

[GM84]   GOLDWASSER, SHAFI and SILVIO MICALI: *Probabilistic Encryption*. J. Comput. Syst. Sci., 28(2):270–299, 1984.

[GPS08]   GALBRAITH, S. D., K. G. PATERSON and N. P. SMART: *Pairings for cryptographers*. Discrete Applied Mathematics, 156(16):3113–3121, 2008.

[GPSW06]   GOYAL, VIPUL, OMKANT PANDEY, AMIT SAHAI and BRENT WATERS: *Attribute-Based Encryption for Fine-Grained Access Control of Encrypted Data*. In JUELS, ARI, REBECCA N. WRIGHT and SABRINA DE CAPITANI DI VIMERCATI (editors): *ACM CCS*

06: *13th Conference on Computer and Communications Security,* pages 89–98. ACM Press, October / November 2006. Available as Cryptology eprint Archive Report 2006/309.

[HLOV09]   HEMENWAY, BRETT, BENOIT LIBERT, RAFAIL OSTROVSKY and DAMIEN VERGNAUD: *Lossy Encryption: Constructions from General Assumptions and Efficient Selective Opening Chosen Ciphertext Security.* Cryptology eprint Archive, Report 2009/088, 2009. http://eprint.iacr.org/.

[Hoh03]   HOHENBERGER, S.: *The Cryptographic Impact of Groups with Infeasible Inversion.* Master's thesis, Massachusetts Institute of Technology, 2003.

[Jag07]   JAGER, TIBOR: *Generic Group Algorithms.* Master's thesis, Ruhr-University Bochum, 2007.

[Jou00]   JOUX, ANTOINE: *A One Round Protocol for Tripartite Diffie-Hellman.* In BOSMA, WIEB (editor): *ANTS,* volume 1838 of *Lecture Notes in Computer Science,* pages 385–394. Springer, 2000.

[Jou04]   JOUX, ANTOINE: *A One Round Protocol for Tripartite Diffie-Hellman.* Journal of Cryptology, 17(4):263–276, September 2004.

[JR10]   JAGER, TIBOR and ANDY RUPP: *The Semi-Generic Group Model and Applications to Pairing-Based Cryptography.* In ABE, MASAYUKI (editor): *Advances in Cryptology – ASIACRYPT 2010,* volume 6477 of *Lecture Notes in Computer Science,* pages 539–556. Springer, December 2010.

[JS08]   JAGER, TIBOR and JÖRG SCHWENK: *On the Equivalence of Generic Group Models.* In BAEK, JOONSANG, FENG BAO, KEFEI CHEN and XUEJIA LAI (editors): *ProvSec,* volume 5324 of *Lecture Notes in Computer Science,* pages 200–209. Springer, 2008.

[JS09]   JAGER, TIBOR and JÖRG SCHWENK: *On the Analysis of Cryptographic Assumptions in the Generic Ring Model.* In MATSUI, MITSURU (editor): *Advances in Cryptology – ASIACRYPT 2009,* volume 5912 of *Lecture Notes in Computer Science,* pages 399–416. Springer, December 2009.

[KG06]   KILTZ, EIKE and DAVID GALINDO: *Direct Chosen-Ciphertext Secure Identity-Based Key Encapsulation Without Random Oracles*. In BATTEN, LYNN MARGARET and REIHANEH SAFAVI-NAINI (editors): *ACISP 06: 11th Australasian Conference on Information Security and Privacy*, volume 4058 of *Lecture Notes in Computer Science*, pages 336–347. Springer, July 2006.

[KM07]   KOBLITZ, NEAL and ALFRED MENEZES: *Another look at generic groups*. Advances in Mathematics of Communications, 1:13–28, 2007.

[KSW08]   KATZ, JONATHAN, AMIT SAHAI and BRENT WATERS: *Predicate Encryption Supporting Disjunctions, Polynomial Equations, and Inner Products*. In SMART, NIGEL P. (editor): *Advances in Cryptology – EUROCRYPT 2008*, volume 4965 of *Lecture Notes in Computer Science*, pages 146–162. Springer, April 2008.

[LOS⁺10] LEWKO, ALLISON B., TATSUAKI OKAMOTO, AMIT SAHAI, KATSUYUKI TAKASHIMA and BRENT WATERS: *Fully Secure Functional Encryption: Attribute-Based Encryption and (Hierarchical) Inner Product Encryption*. In GILBERT, HENRI (editor): *Advances in Cryptology – EUROCRYPT 2010*, volume 6110 of *Lecture Notes in Computer Science*, pages 62–91. Springer, May 2010.

[LR06]   LEANDER, GREGOR and ANDY RUPP: *On the Equivalence of RSA and Factoring Regarding Generic Ring Algorithms*. In LAI, XUEJIA and KEFEI CHEN (editors): *Advances in Cryptology – ASIACRYPT 2006*, volume 4284 of *Lecture Notes in Computer Science*, pages 241–251. Springer, December 2006.

[Mau94]   MAURER, UELI M.: *Towards the Equivalence of Breaking the Diffie-Hellman Protocol and Computing Discrete Logarithms*. In DESMEDT, YVO (editor): *Advances in Cryptology – CRYPTO'94*, volume 839 of *Lecture Notes in Computer Science*, pages 271–281. Springer, August 1994.

[Mau05]   MAURER, UELI M.: *Abstract Models of Computation in Cryptography*. In SMART, NIGEL P. (editor): *IMA Int. Conf.*, volume 3796 of *Lecture Notes in Computer Science*, pages 1–12. Springer, 2005.

[May04]   MAY, ALEXANDER: *Computing the RSA Secret Key Is Deterministic Polynomial Time Equivalent to Factoring*. In FRANKLIN,

MATTHEW (editor): *Advances in Cryptology – CRYPTO 2004*, volume 3152 of *Lecture Notes in Computer Science*, pages 213–219. Springer, August 2004.

[Mic05]   MICCIANCIO, DANIELE: *The RSA Group is Pseudo-Free*. In CRAMER, RONALD (editor): *Advances in Cryptology – EURO-CRYPT 2005*, volume 3494 of *Lecture Notes in Computer Science*, pages 387–403. Springer, May 2005.

[Mil76]   MILLER, GARY L.: *Riemann's Hypothesis and Tests for Primality*. J. Comput. Syst. Sci., 13(3):300–317, 1976.

[MOV93]   MENEZES, A., T. OKAMOTO and S. VANSTONE: *Reducing elliptic curve logarithms to logarithms in a finite field*. IEEE Transactions on Information Theory, 39(5):1639–1646, 1993.

[MR07]   MAURER, UELI M. and DOMINIK RAUB: *Black-Box Extension Fields and the Inexistence of Field-Homomorphic One-Way Permutations*. In KUROSAWA, KAORU (editor): *Advances in Cryptology – ASIACRYPT 2007*, volume 4833 of *Lecture Notes in Computer Science*, pages 427–443. Springer, December 2007.

[MW96]   MAURER, UELI M. and STEFAN WOLF: *Diffie-Hellman Oracles*. In KOBLITZ, NEAL (editor): *Advances in Cryptology – CRYPTO'96*, volume 1109 of *Lecture Notes in Computer Science*, pages 268–282. Springer, August 1996.

[MW98]   MAURER, UELI M. and STEFAN WOLF: *Lower Bounds on Generic Algorithms in Groups*. In NYBERG, KAISA (editor): *Advances in Cryptology – EUROCRYPT'98*, volume 1403 of *Lecture Notes in Computer Science*, pages 72–84. Springer, May / June 1998.

[MW99]   MAURER, UELI M. and STEFAN WOLF: *The Relationship Between Breaking the Diffie-Hellman Protocol and Computing Discrete Logarithms*. SIAM J. Comput., 28(5):1689–1721, 1999.

[Nec94]   NECHAEV, V. I.: *Complexity of a Determinate Algorithm for the Discrete Logarithm*. Mathematical Notes, 55(2):165–172, 1994.

[OSW07]   OSTROVSKY, RAFAIL, AMIT SAHAI and BRENT WATERS: *Attribute-based encryption with non-monotonic access structures*. In NING, PENG, SABRINA DE CAPITANI DI VIMERCATI and

PAUL F. SYVERSON (editors): *ACM CCS 07: 14th Conference on Computer and Communications Security*, pages 195–203. ACM Press, October 2007.

[OT10]   OKAMOTO, TATSUAKI and KATSUYUKI TAKASHIMA: *Fully Secure Functional Encryption with General Relations from the Decisional Linear Assumption*. In RABIN, TAL (editor): *Advances in Cryptology – CRYPTO 2010*, volume 6223 of *Lecture Notes in Computer Science*, pages 191–208. Springer, August 2010.

[Pai99]   PAILLIER, PASCAL: *Public-Key Cryptosystems Based on Composite Degree Residuosity Classes*. In STERN, JACQUES (editor): *Advances in Cryptology – EUROCRYPT'99*, volume 1592 of *Lecture Notes in Computer Science*, pages 223–238. Springer, May 1999.

[Pol75]   POLLARD, JOHN M.: *A Monte Carlo method for factorization*. BIT, 15:331–334, 1975.

[Rab79]   RABIN, MICHAEL O.: *Digitalized signatures and public-key functions as intractable as factorization*. MIT Laboratory for Computer Science, 1979.

[Riv04]   RIVEST, RONALD L.: *On the Notion of Pseudo-Free Groups*. In NAOR, MONI (editor): *TCC 2004: 1st Theory of Cryptography Conference*, volume 2951 of *Lecture Notes in Computer Science*, pages 505–521. Springer, February 2004.

[RLB+08]   RUPP, ANDY, GREGOR LEANDER, ENDRE BANGERTER, ALEXANDER W. DENT and AHMAD-REZA SADEGHI: *Sufficient Conditions for Intractability over Black-Box Groups: Generic Lower Bounds for Generalized DL and DH Problems*. In PIEPRZYK, JOSEF (editor): *Advances in Cryptology – ASIACRYPT 2008*, volume 5350 of *Lecture Notes in Computer Science*, pages 489–505. Springer, December 2008.

[RS08]   ROSEN, ALON and GIL SEGEV: *Efficient Lossy Trapdoor Functions based on the Composite Residuosity Assumption*. Cryptology ePrint Archive, Report 2008/134, 2008. http://eprint.iacr.org/.

[RSA78]   RIVEST, RONALD L., ADI SHAMIR and LEONARD M. ADLEMAN: *A method for obtaining digital signatures and public-key cryptosystems*. Communications of the ACM, 21:120–126, 1978.

[Sha71]   SHANKS, DANIEL: *Class number, a theory of factorization, and genera.* pages 415–440, 1971.

[Sho97]   SHOUP, VICTOR: *Lower Bounds for Discrete Logarithms and Related Problems.* In FUMY, WALTER (editor): *Advances in Cryptology – EUROCRYPT'97,* volume 1233 of *Lecture Notes in Computer Science,* pages 256–266. Springer, May 1997.

[Sho04]   SHOUP, VICTOR: *Sequences of games: a tool for taming complexity in security proofs.* Cryptology ePrint Archive, Report 2004/332, 2004. http://eprint.iacr.org/.

[Sho08]   SHOUP, VICTOR: *A Computational Introduction to Number Theory and Algebra.* Cambridge University Press, Second edition, 2008.

[Wat05]   WATERS, BRENT R.: *Efficient Identity-Based Encryption Without Random Oracles.* In CRAMER, RONALD (editor): *Advances in Cryptology – EUROCRYPT 2005,* volume 3494 of *Lecture Notes in Computer Science,* pages 114–127. Springer, May 2005.

[Wol99]   WOLF, STEFAN: *Information-theoretically and computationally secure key agreement in cryptography.* PhD thesis, ETH Zurich, 1999. ETH dissertation No. 13138.

GPSR Compliance
The European Union's (EU) General Product Safety Regulation (GPSR) is a set
of rules that requires consumer products to be safe and our obligations to
ensure this.

If you have any concerns about our products, you can contact us on

ProductSafety@springernature.com

In case Publisher is established outside the EU, the EU authorized
representative is:

Springer Nature Customer Service Center GmbH
Europaplatz 3
69115 Heidelberg, Germany

www.ingramcontent.com/pod-product-compliance
Lightning Source LLC
LaVergne TN
LVHW052307060326
832902LV00021B/3753

*9 7 8 3 8 3 4 8 1 9 8 9 5 *